Science and Fiction

Science and Fiction – A Springer Series

This collection of entertaining and thought-provoking books will appeal equally to science buffs, scientists and science-fiction fans. It was born out of the recognition that scientific discovery and the creation of plausible fictional scenarios are often two sides of the same coin. Each relies on an understanding of the way the world works, coupled with the imaginative ability to invent new or alternative explanations—and even other worlds. Authored by practicing scientists as well as writers of hard science fiction, these books explore and exploit the borderlands between accepted science and its fictional counterpart. Uncovering mutual influences, promoting fruitful interaction, narrating and analyzing fictional scenarios, together they serve as a reaction vessel for inspired new ideas in science, technology, and beyond.

Whether fiction, fact, or forever undecidable: the Springer Series "Science and Fiction" intends to go where no one has gone before!

Its largely non-technical books take several different approaches. Journey with their authors as they

- Indulge in science speculation – describing intriguing, plausible yet unproven ideas;
- Exploit science fiction for educational purposes and as a means of promoting critical thinking;
- Explore the interplay of science and science fiction – throughout the history of the genre and looking ahead;
- Delve into related topics including, but not limited to: science as a creative process, the limits of science, interplay of literature and knowledge;
- Tell fictional short stories built around well-defined scientific ideas, with a supplement summarizing the science underlying the plot.

Readers can look forward to a broad range of topics, as intriguing as they are important. Here just a few by way of illustration:

- Time travel, superluminal travel, wormholes, teleportation
- Extraterrestrial intelligence and alien civilizations
- Artificial intelligence, planetary brains, the universe as a computer, simulated worlds
- Non-anthropocentric viewpoints
- Synthetic biology, genetic engineering, developing nanotechnologies
- Eco/infrastructure/meteorite-impact disaster scenarios
- Future scenarios, transhumanism, posthumanism, intelligence explosion
- Virtual worlds, cyberspace dramas
- Consciousness and mind manipulation

More information about this series at http://www.springer.com/series/11657

Stephen Webb

New Light Through Old Windows: Exploring Contemporary Science Through 12 Classic Science Fiction Tales

 Springer

Stephen Webb
DCQE University of Portsmouth
Lee-on-the-Solent, UK

ISSN 2197-1188   ISSN 2197-1196   (electronic)
Science and Fiction
ISBN 978-3-030-03194-7   ISBN 978-3-030-03195-4   (eBook)
https://doi.org/10.1007/978-3-030-03195-4

Library of Congress Control Number: 2018963833

Cover illustration: Science fiction illustration of the view through an archway in the old town across the bay to the modern buildings of the future city on a bright sunny day, 3d digitally rendered illustration.
Cover design by Algol/shutterstock.com

This Springer imprint is published by the registered company Springer Nature Switzerland AG
The registered company address is: Gewerbestrasse 11, 6330 Cham, Switzerland

To my Aunt Edna

Preface

As background research for my book *All the Wonder that Would Be* I set myself the happy task of rereading some of my favourite SF stories. Most of them were first published in the period 1935–1985, but upon rereading them it soon became clear to me the extent to which their authors had been standing on the shoulders of earlier giants such as H.G. Wells and Arthur Conan Doyle (or at least on the shoulders of the very tall, such as Ambrose Bierce and William Hope Hodgson). And reading those earlier tales led me to meet some classic SF for the first time—I hadn't realized, for example, how many key science fictional themes E.P. Mitchell seems to have pioneered.

Once you get past their outdated social views and sometimes unfashionable phraseology, these old stories can be rather rewarding. Their authors took the accepted knowledge of the day as a springboard for extrapolation. In other words, they engaged in the same activity that occupies modern-day SF writers—and it's fascinating to compare the results of their extrapolations with our current understanding of science. In some cases, the authors were remarkably prescient; in other cases, they were right for the wrong reasons; and in some cases, they were just plain wrong. In all cases, though, I believe we can learn from them when we attempt to make our own extrapolations and imagine what the future might hold for us.

This book collects a dozen SF stories which, although they were published a long time ago (the oldest story appeared in 1817 and the youngest in 1934), might nevertheless be unfamiliar to younger SF fans. Each story is accompanied by a short scientific commentary. The intention of the commentaries is not to provide full coverage of the relevant science—that would make the book far too long. Rather, they simply contain some thoughts about contemporary science sparked by looking at these old gems. For those who would like to delve deeper into the science, a further reading section at the end of each commentary should provide sufficient pointers to the relevant modern literature.

Lee-on-the-Solent, UK   Stephen Webb

Acknowledgements

I would like to thank Chris Caron, not only for his wise advice but also for passing on numerous comments and ideas that helped me greatly.

Joey Meyer and Professor Steve Ruff kindly allowed me to reproduce images, and I appreciate their quick response to my requests.

Despite many years of struggle with the German language, my linguistic skills remain at a level that lets me order a beer but do little else. So I thank my wife Heike for help with Chap. 7 and for her patience in general. Jessica, as always, is an inspiration.

Contents

1

Life … But Not as We Know It

The Terror of Blue John Gap (Arthur Conan Doyle)

The following narrative was found among the papers of Dr. James Hardcastle, who died of phthisis on February 4th, 1908, at 36, Upper Coventry Flats, South Kensington. Those who knew him best, while refusing to express an opinion upon this particular statement, are unanimous in asserting that he was a man of a sober and scientific turn of mind, absolutely devoid of imagination, and most unlikely to invent any abnormal series of events. The paper was contained in an envelope, which was docketed, "A Short Account of the Circumstances which occurred near Miss Allerton's Farm in North-West Derbyshire in the Spring of Last Year." The envelope was sealed, and on the other side was written in pencil—

DEAR SEATON,—

It may interest, and perhaps pain you, to know that the incredulity with which you met my story has prevented me from ever opening my mouth upon the subject again. I leave this record after my death, and perhaps strangers may be found to have more confidence in me than my friend.

Inquiry has failed to elicit who this Seaton may have been. I may add that the visit of the deceased to Allerton's Farm, and the general nature of the alarm there, apart from his particular explanation, have been absolutely established. With this foreword I append his account exactly as he left it. It is in the

© Springer Nature Switzerland AG 2019
S. Webb, *New Light Through Old Windows: Exploring Contemporary Science Through 12 Classic Science Fiction Tales*, Science and Fiction,
https://doi.org/10.1007/978-3-030-03195-4_1

form of a diary, some entries in which have been expanded, while a few have been erased.

April 17.—Already I feel the benefit of this wonderful upland air. The farm of the Allertons lies fourteen hundred and twenty feet above sea-level, so it may well be a bracing climate. Beyond the usual morning cough I have very little discomfort, and, what with the fresh milk and the homegrown mutton, I have every chance of putting on weight. I think Saunderson will be pleased.

The two Miss Allertons are charmingly quaint and kind, two dear little hard-working old maids, who are ready to lavish all the heart which might have gone out to husband and to children upon an invalid stranger. Truly, the old maid is a most useful person, one of the reserve forces of the community. They talk of the superfluous woman, but what would the poor superfluous man do without her kindly presence? By the way, in their simplicity they very quickly let out the reason why Saunderson recommended their farm. The Professor rose from the ranks himself, and I believe that in his youth he was not above scaring crows in these very fields.

It is a most lonely spot, and the walks are picturesque in the extreme. The farm consists of grazing land lying at the bottom of an irregular valley. On each side are the fantastic limestone hills, formed of rock so soft that you can break it away with your hands. All this country is hollow. Could you strike it with some gigantic hammer it would boom like a drum, or possibly cave in altogether and expose some huge subterranean sea. A great sea there must surely be, for on all sides the streams run into the mountain itself, never to reappear. There are gaps everywhere amid the rocks, and when you pass through them you find yourself in great caverns, which wind down into the bowels of the earth. I have a small bicycle lamp, and it is a perpetual joy to me to carry it into these weird solitudes, and to see the wonderful silver and black effect when I throw its light upon the stalactites which drape the lofty roofs. Shut off the lamp, and you are in the blackest darkness. Turn it on, and it is a scene from the Arabian Nights.

But there is one of these strange openings in the earth which has a special interest, for it is the handiwork, not of nature, but of man. I had never heard of Blue John when I came to these parts. It is the name given to a peculiar mineral of a beautiful purple shade, which is only found at one or two places in the world. It is so rare that an ordinary vase of Blue John would be valued at a great price. The Romans, with that extraordinary instinct of theirs, discovered that it was to be found in this valley, and sank a horizontal shaft deep into the mountain side. The opening of their mine has been called Blue John Gap, a clean-cut arch in the rock, the mouth all overgrown with bushes. It is a goodly passage which the Roman miners have cut, and it intersects some of the great water-worn caves, so that if you enter Blue John Gap you would do well to mark your steps and to have a good store of candles, or you may never make your way back to the daylight again. I have not yet gone deeply into it, but this very day I stood at the mouth of the arched tunnel, and peering down into the black recesses beyond, I vowed that when my health returned I would devote some holiday to exploring those mysterious depths and finding out for myself how far the Roman had penetrated into the Derbyshire hills.

Strange how superstitious these countrymen are! I should have thought better of young Armitage, for he is a man of some education and character, and a very fine fellow for his station in life. I was standing at the Blue John Gap when he came across the field to me.

"Well, doctor," said he, "you're not afraid, anyhow."

"Afraid!" I answered. "Afraid of what?"

"Of it," said he, with a jerk of his thumb towards the black vault, "of the Terror that lives in the Blue John Cave."

How absurdly easy it is for a legend to arise in a lonely countryside! I examined him as to the reasons for his weird belief. It seems that from time to time sheep have been missing from the fields, carried bodily away, according to Armitage. That they could have wandered away of their own accord and disappeared among the mountains was an explanation to which he would not listen. On one occasion a pool of blood had been found, and some tufts of wool. That also, I pointed out, could be explained in a perfectly natural way. Further, the nights upon which sheep disappeared were invariably very dark, cloudy nights with no moon. This I met with the obvious retort that those were the nights which a commonplace sheep-stealer would naturally choose for his work. On one occasion a gap had been made in a wall, and some of the stones scattered for a considerable distance. Human agency again, in my opinion. Finally, Armitage clinched all his arguments by telling me that he had actually heard the Creature—indeed, that anyone could hear it who remained long enough at the Gap. It was a distant roaring of an immense volume. I could not but smile at this, knowing, as I do, the strange reverberations which come out of an underground water system running amid the chasms of a limestone formation. My incredulity annoyed Armitage so that he turned and left me with some abruptness.

And now comes the queer point about the whole business. I was still standing near the mouth of the cave turning over in my mind the various statements of Armitage, and reflecting how readily they could be explained away, when suddenly, from the depth of the tunnel beside me, there issued a most extraordinary sound. How shall I describe it? First of all, it seemed to be a great distance away, far down in the bowels of the earth. Secondly, in spite of this suggestion of distance, it was very loud. Lastly, it was not a boom, nor a crash, such as one would associate with falling water or tumbling rock, but it was a high whine, tremulous and vibrating, almost like the whinnying of a horse. It was certainly a most remarkable experience, and one which for a moment, I must admit, gave a new significance to Armitage's words. I waited by the Blue John Gap for half an hour or more, but there was no return of the sound, so at last I wandered back to the farmhouse, rather mystified by what had occurred. Decidedly I shall explore that cavern when my strength is restored. Of course, Armitage's explanation is too absurd for discussion, and yet that sound was certainly very strange. It still rings in my ears as I write.

April 20.—In the last three days I have made several expeditions to the Blue John Gap, and have even penetrated some short distance, but my bicycle lantern is so small and weak that I dare not trust myself very far. I shall do the thing more systematically. I have heard no sound at all, and could almost believe that I had been the victim of some hallucination suggested, perhaps, by Armitage's conversation. Of course, the whole idea is absurd, and yet I must confess that those bushes at the entrance of the cave do present an appearance as if some heavy creature had forced its way through them. I begin to be keenly interested. I have said nothing to the Miss Allertons, for they are quite superstitious enough already, but I have bought some candles, and mean to investigate for myself.

I observed this morning that among the numerous tufts of sheep's wool which lay among the bushes near the cavern there was one which was smeared with blood. Of course, my reason tells me that if sheep wander into such rocky places they are likely to injure themselves, and yet somehow that splash of crimson gave me a sudden shock, and for a moment I found myself shrinking back in horror from the old Roman arch. A fetid breath seemed to ooze from the black depths into which I peered. Could it indeed be possible that some nameless thing, some dreadful presence, was lurking down yonder? I should have been incapable of such feelings in the days of my strength, but one grows more nervous and fanciful when one's health is shaken.

For the moment I weakened in my resolution, and was ready to leave the secret of the old mine, if one exists, for ever unsolved. But tonight my interest has returned and my nerves grown more steady. Tomorrow I trust that I shall have gone more deeply into this matter.

April 22.—Let me try and set down as accurately as I can my extraordinary experience of yesterday. I started in the afternoon, and made my way to the Blue John Gap. I confess that my misgivings returned as I gazed into its depths, and I wished that I had brought a companion to share my exploration.

Finally, with a return of resolution, I lit my candle, pushed my way through the briars, and descended into the rocky shaft.

It went down at an acute angle for some fifty feet, the floor being covered with broken stone. Thence there extended a long, straight passage cut in the solid rock. I am no geologist, but the lining of this corridor was certainly of some harder material than limestone, for there were points where I could actually see the tool-marks which the old miners had left in their excavation, as fresh as if they had been done yesterday. Down this strange, old-world corridor I stumbled, my feeble flame throwing a dim circle of light around me, which made the shadows beyond the more threatening and obscure. Finally, I came to a spot where the Roman tunnel opened into a water-worn cavern—a huge hall, hung with long white icicles of lime deposit. From this central chamber I could dimly perceive that a number of passages worn by the subterranean streams wound away into the depths of the earth. I was standing there wondering whether I had better return, or whether I dare venture farther into this dangerous labyrinth, when my eyes fell upon something at my feet which strongly arrested my attention.

The greater part of the floor of the cavern was covered with boulders of rock or with hard incrustations of lime, but at this particular point there had been a drip from the distant roof, which had left a patch of soft mud. In the very centre of this there was a huge mark—an ill-defined blotch, deep, broad and irregular, as if a great boulder had fallen upon it. No loose stone lay near, however, nor was there anything to account for the impression. It was far too large to be caused by any possible animal, and besides, there was only the one, and the patch of mud was of such a size that no reasonable stride could have covered it. As I rose from the examination of that singular mark and then looked round into the black shadows which hemmed me in, I must confess that I felt for a moment a most unpleasant sinking of my heart, and that, do what I could, the candle trembled in my outstretched hand.

I soon recovered my nerve, however, when I reflected how absurd it was to associate so huge and shapeless a mark with the track of any known animal. Even an elephant could not have produced it. I determined, therefore, that I would not be scared by vague and senseless fears from carrying out my exploration. Before proceeding, I took good note of a curious rock formation in the wall by which I could recognize the entrance of the Roman tunnel. The precaution was very necessary, for the great cave, so far as I could see it, was intersected by passages. Having made sure of my position, and reassured myself by examining my spare candles and my matches, I advanced slowly over the rocky and uneven surface of the cavern.

And now I come to the point where I met with such sudden and desperate disaster. A stream, some twenty feet broad, ran across my path, and I walked for some little distance along the bank to find a spot where I could cross dry-shod. Finally, I came to a place where a single flat boulder lay near the centre, which I could reach in a stride. As it chanced, however, the rock had been cut away and made top-heavy by the rush of the stream, so that it tilted over as I landed on it and shot me into the ice-cold water. My candle went out, and I found myself floundering about in utter and absolute darkness.

I staggered to my feet again, more amused than alarmed by my adventure. The candle had fallen from my hand, and was lost in the stream, but I had two others in my pocket, so that it was of no importance. I got one of them ready, and drew out my box of matches to light it. Only then did I realize my position. The box had been soaked in my fall into the river. It was impossible to strike the matches.

A cold hand seemed to close round my heart as I realized my position. The darkness was opaque and horrible. It was so utter one put one's hand up to one's face as if to press off something solid. I stood still, and by an effort I steadied myself. I tried to reconstruct in my mind a map of the floor of the cavern as I had last seen it. Alas! the bearings which had impressed themselves upon my mind were high on the wall, and not to be found by touch. Still, I remembered in a general way how the sides were situated, and I hoped that by groping my way along them I should at last come to the opening of the Roman tunnel. Moving very slowly, and continually striking against the rocks, I set out on this desperate quest.

But I very soon realized how impossible it was. In that black, velvety darkness one lost all one's bearings in an instant. Before I had made a dozen paces, I was utterly bewildered as to my whereabouts. The rippling of the stream, which was the one sound audible, showed me where it lay, but the moment that I left its bank I was utterly lost. The idea of finding my way back in absolute darkness through that limestone labyrinth was clearly an impossible one. I sat down upon a boulder and reflected upon my unfortunate plight. I had not told anyone that I proposed to come to the Blue John mine, and it was unlikely that a search party would come after me. Therefore I must trust to my own resources to get clear of the danger. There was only one hope, and that was that the matches might dry. When I fell into the river, only half of me had got thoroughly wet. My left shoulder had remained above the water. I took the box of matches, therefore, and put it into my left armpit. The moist air of the cavern might possibly be counteracted by the heat of my body, but even so, I knew that I could not hope to get a light for many hours. Meanwhile there was nothing for it but to wait.

By good luck I had slipped several biscuits into my pocket before I left the farmhouse. These I now devoured, and washed them down with a draught from that wretched stream which had been the cause of all my misfortunes. Then I felt about for a comfortable seat among the rocks, and, having discovered a place where I could get a support for my back, I stretched out my legs and settled myself down to wait. I was wretchedly damp and cold, but I tried to cheer myself with the reflection that modern science prescribed open windows and walks in all weather for my disease. Gradually, lulled by the monotonous gurgle of the stream, and by the absolute darkness, I sank into an uneasy slumber.

How long this lasted I cannot say. It may have been for an hour, it may have been for several. Suddenly I sat up on my rock couch, with every nerve thrilling and every sense acutely on the alert. Beyond all doubt I had heard a sound—some sound very distinct from the gurgling of the waters. It had passed, but the reverberation of it still lingered in my ear. Was it a search party? They would most certainly have shouted, and vague as this sound was which had wakened me, it was very distinct from the human voice. I sat palpitating and hardly daring to breathe. There it was again! And again! Now it had become continuous. It was a tread—yes, surely it was the tread of some living creature. But what a tread it was! It gave one the impression of enormous weight carried upon sponge-like feet, which gave forth a muffled but ear-filling sound. The darkness was as complete as ever, but the tread was regular and decisive. And it was coming beyond all question in my direction. My skin grew cold, and my hair stood on end as I listened to that steady and ponderous footfall. There was some creature there, and surely by the speed of its advance, it was one which could see in the dark. I crouched low on my rock and tried to blend myself into it. The steps grew nearer still, then stopped, and presently I was aware of a loud lapping and gurgling. The creature was drinking at the stream. Then again there was silence, broken by a succession of long sniffs and snorts of tremendous volume and energy. Had it caught the scent of me? My own nostrils were filled by a low fetid odour, mephitic and abominable. Then I heard the steps again. They were on my side of the stream now. The stones rattled within a few yards of where I lay. Hardly daring to breathe, I crouched upon my rock. Then the steps drew away. I heard the splash as it returned across the river, and the sound died away into the distance in the direction from which it had come.

For a long time I lay upon the rock, too much horrified to move. I thought of the sound which I had heard coming from the depths of the cave, of Armitage's fears, of the strange impression in the mud, and now came this final and absolute proof that there was indeed some inconceivable monster, something utterly unearthly and dreadful, which lurked in the hollow of the mountain. Of its nature or form I could frame no conception, save that it was both light-footed and gigantic. The combat between my reason, which told me that such things could not be, and my senses, which told me that they were, raged within me as I lay. Finally, I was almost ready to persuade myself that this experience had been part of some evil dream, and that my abnormal condition might have conjured up an hallucination. But there remained one final experience which removed the last possibility of doubt from my mind. I had taken my matches from my armpit and felt them. They seemed perfectly hard and dry. Stooping down into a crevice of the rocks, I tried one of them. To my delight it took fire at once. I lit the candle, and, with a terrified backward glance into the obscure depths of the cavern, I hurried in the direction of the Roman passage. As I did so I passed the patch of mud on which I had seen the huge imprint. Now I stood astonished before it, for there were three similar imprints upon its surface, enormous in size, irregular in outline, of a depth which indicated the ponderous weight which had left them. Then a great terror surged over me. Stooping and shading my candle with my hand, I ran in a frenzy of fear to the rocky archway, hastened up it, and never stopped until, with weary feet and panting lungs, I rushed up the final slope of stones, broke through the tangle of briars, and flung myself exhausted upon the soft grass under the peaceful light of the stars. It was three in the morning when I reached the farmhouse, and today I am all unstrung and quivering after my terrific adventure. As yet I have told no one. I must move warily in the matter. What would the poor lonely women, or the uneducated yokels here think of it if I were to tell them my experience? Let me go to someone who can understand and advise.

April 25.—I was laid up in bed for two days after my incredible adventure in the cavern. I use the adjective with a very definite meaning, for I have had an experience since which has shocked me almost as much as the other. I have said that I was looking round for someone who could advise me. There is a Dr. Mark Johnson who practices some few miles away, to whom I had a note of recommendation from Professor Saunderson. To him I drove, when I was strong enough to get about, and I recounted to him my whole strange experience. He listened intently, and then carefully examined me, paying special attention to my reflexes and to the pupils of my eyes. When he had finished, he refused to discuss my adventure, saying that it was entirely beyond him, but he gave me the card of a Mr. Picton at Castleton, with the advice that I should instantly go to him and tell him the story exactly as I had done to himself. He was, according to my adviser, the very man who was pre-eminently suited to help me. I went on to the station, therefore, and made my way to the little town, which is some ten miles away. Mr. Picton appeared to be a man of importance, as his brass plate was displayed upon the door of a considerable building on the outskirts of the town. I was about to ring his bell, when some misgiving came into my mind, and, crossing to a neighbouring shop, I asked the man behind the counter if he could tell me anything of Mr. Picton. "Why," said he, "he is the best mad doctor in Derbyshire, and yonder is his asylum." You can imagine that it was not long before I had shaken the dust of Castleton from my feet and returned to the farm, cursing all unimaginative pedants who cannot conceive that there may be things in creation which have never yet chanced to come across their mole's vision. After all, now that I am cooler, I can afford to admit that I have been no more sympathetic to Armitage than Dr. Johnson has been to me.

April 27. When I was a student I had the reputation of being a man of courage and enterprise. I remember that when there was a ghost-hunt at Coltbridge it was I who sat up in the haunted house. Is it advancing years (after all, I am only thirty-five), or is it this physical malady which has caused degeneration? Certainly my heart quails when I think of that horrible cavern in the hill, and the certainty that it has some monstrous occupant. What shall I do? There is not an hour in the day that I do not debate the question. If I say nothing, then the mystery remains unsolved. If I do say anything, then I have the alternative of mad alarm over the whole countryside, or of absolute incredulity which may end in consigning me to an asylum. On the whole, I think that my best course is to wait, and to prepare for some expedition which shall be more deliberate and better thought out than the last. As a first step I have been to Castleton and obtained a few essentials—a large acetylene lantern for one thing, and a good double-barrelled sporting rifle for another. The latter I have hired, but I have bought a dozen heavy game cartridges, which would bring down a rhinoceros. Now I am ready for my troglodyte friend. Give me better health and a little spate of energy, and I shall try conclusions with him yet. But who and what is he? Ah! there is the question which stands between me and my sleep. How many theories do I form, only to discard each in turn! It is all so utterly unthinkable. And yet the cry, the footmark, the tread in the cavern—no reasoning can get past these I think of the old-world legends of dragons and of other monsters. Were they, perhaps, not such fairy-tales as we have thought? Can it be that there is some fact which underlies them, and am I, of all mortals, the one who is chosen to expose it?

May 3.—For several days I have been laid up by the vagaries of an English spring, and during those days there have been developments, the true and sinister meaning of which no one can appreciate save myself. I may say that we have had cloudy and moonless nights of late, which according to my information were the seasons upon which sheep disappeared. Well, sheep have disappeared. Two of Miss Allerton's, one of old Pearson's of the Cat Walk, and one of Mrs. Moulton's. Four in all during three nights. No trace is left of them at all, and the countryside is buzzing with rumours of gipsies and of sheep-stealers.

But there is something more serious than that. Young Armitage has disappeared also. He left his moorland cottage early on Wednesday night and has never been heard of since. He was an unattached man, so there is less sensation than would otherwise be the case. The popular explanation is that he owes money, and has found a situation in some other part of the country, whence he will presently write for his belongings. But I have grave misgivings. Is it not much more likely that the recent tragedy of the sheep has caused him to take some steps which may have ended in his own destruction? He may, for example, have lain in wait for the creature and been carried off by it into the recesses of the mountains. What an inconceivable fate for a civilized Englishman of the twentieth century! And yet I feel that it is possible and even probable. But in that case, how far am I answerable both for his death and for any other mishap which may occur? Surely with the knowledge I already possess it must be my duty to see that something is done, or if necessary to do it myself. It must be the latter, for this morning I went down to the local police-station and told my story. The inspector entered it all in a large book and bowed me out with commendable gravity, but I heard a burst of laughter before I had got down his garden path. No doubt he was recounting my adventure to his family.

June 10.—I am writing this, propped up in bed, six weeks after my last entry in this journal. I have gone through a terrible shock both to mind and body, arising from such an experience as has seldom befallen a human being before. But I have attained my end. The danger from the Terror which dwells in the Blue John Gap has passed never to return. Thus much at least I, a broken invalid, have done for the common good. Let me now recount what occurred as clearly as I may.

The night of Friday, May 3rd, was dark and cloudy—the very night for the monster to walk. About eleven o'clock I went from the farmhouse with my lantern and my rifle, having first left a note upon the table of my bedroom in which I said that, if I were missing, search should be made for me in the direction of the Gap. I made my way to the mouth of the Roman shaft, and, having perched myself among the rocks close to the opening, I shut off my lantern and waited patiently with my loaded rifle ready to my hand.

It was a melancholy vigil. All down the winding valley I could see the scattered lights of the farmhouses, and the church clock of Chapel-le-Dale tolling the hours came faintly to my ears. These tokens of my fellow-men served only to make my own position seem the more lonely, and to call for a greater effort to overcome the terror which tempted me continually to get back to the farm, and abandon for ever this dangerous quest. And yet there lies deep in every man a rooted self-respect which makes it hard for him to turn back from that which he has once undertaken. This feeling of personal pride was my salvation now, and it was that alone which held me fast when every instinct of my nature was dragging me away. I am glad now that I had the strength. In spite of all that is has cost me, my manhood is at least above reproach.

Twelve o'clock struck in the distant church, then one, then two. It was the darkest hour of the night. The clouds were drifting low, and there was not a star in the sky. An owl was hooting somewhere among the rocks, but no other sound, save the gentle sough of the wind, came to my ears. And then suddenly I heard it! From far away down the tunnel came those muffled steps, so soft and yet so ponderous. I heard also the rattle of stones as they gave way under that giant tread. They drew nearer. They were close upon me. I heard the crashing of the bushes round the entrance, and then dimly through the darkness I was conscious of the loom of some enormous shape, some monstrous inchoate creature, passing swiftly and very silently out from the tunnel. I was paralysed with fear and amazement. Long as I had waited, now that it had actually come I was unprepared for the shock. I lay motionless and breathless, whilst the great dark mass whisked by me and was swallowed up in the night.

But now I nerved myself for its return. No sound came from the sleeping countryside to tell of the horror which was loose. In no way could I judge how far off it was, what it was doing, or when it might be back. But not a second time should my nerve fail me, not a second time should it pass unchallenged. I swore it between my clenched teeth as I laid my cocked rifle across the rock. And yet it nearly happened. There was no warning of approach now as the creature passed over the grass. Suddenly, like a dark, drifting shadow, the huge bulk loomed up once more before me, making for the entrance of the cave. Again came that paralysis of volition which held my crooked forefinger impotent upon the trigger. But with a desperate effort I shook it off. Even as the brushwood rustled, and the monstrous beast blended with the shadow of the Gap, I fired at the retreating form. In the blaze of the gun I caught a glimpse of a great shaggy mass, something with rough and bristling hair of a withered grey colour, fading away to white in its lower parts, the huge body supported upon short, thick, curving legs. I had just that glance, and then I heard the rattle of the stones as the creature tore down into its burrow. In an instant, with a triumphant revulsion of feeling, I had cast my fears to the wind, and uncovering my powerful lantern, with my rifle in my hand, I sprang down from my rock and rushed after the monster down the old Roman shaft.

My splendid lamp cast a brilliant flood of vivid light in front of me, very different from the yellow glimmer which had aided me down the same passage only twelve days before. As I ran, I saw the great beast lurching along before me, its huge bulk filling up the whole space from wall to wall. Its hair looked like coarse faded oakum, and hung down in long, dense masses which swayed as it moved. It was like an enormous unclipped sheep in its fleece, but in size it was far larger than the largest elephant, and its breadth seemed to be nearly as great as its height. It fills me with amazement now to think that I should have dared to follow such a horror into the bowels of the earth, but when one's blood is up, and when one's quarry seems to be flying, the old primeval hunting-spirit awakes and prudence is cast to the wind. Rifle in hand, I ran at the top of my speed upon the trail of the monster.

I had seen that the creature was swift. Now I was to find out to my cost that it was also very cunning. I had imagined that it was in panic flight, and that I had only to pursue it. The idea that it might turn upon me never entered my excited brain. I have already explained that the passage down which I was racing opened into a great central cave. Into this I rushed, fearful lest I should lose all trace of the beast. But he had turned upon his own traces, and in a moment we were face to face.

That picture, seen in the brilliant white light of the lantern, is etched for ever upon my brain. He had reared up on his hind legs as a bear would do, and stood above me, enormous, menacing—such a creature as no nightmare had ever brought to my imagination. I have said that he reared like a bear, and there was something bear-like—if one could conceive a bear which was ten-fold the bulk of any bear seen upon earth—in his whole pose and attitude, in his great crooked forelegs with their ivory-white claws, in his rugged skin, and in his red, gaping mouth, fringed with monstrous fangs. Only in one point did he differ from the bear, or from any other creature which walks the earth, and even at that supreme moment a shudder of horror passed over me as I observed that the eyes which glistened in the glow of my lantern were huge, projecting bulbs, white and sightless. For a moment his great paws swung over my head. The next he fell forward upon me, I and my broken lantern crashed to the earth, and I remember no more.

When I came to myself I was back in the farmhouse of the Allertons. Two days had passed since my terrible adventure in the Blue John Gap. It seems that I had lain all night in the cave insensible from concussion of the brain, with my left arm and two ribs badly fractured. In the morning my note had been found, a search party of a dozen farmers assembled, and I had been tracked down and carried back to my bedroom, where I had lain in high delirium ever since. There was, it seems, no sign of the creature, and no bloodstain which would show that my bullet had found him as he passed. Save for my own plight and the marks upon the mud, there was nothing to prove that what I said was true.

Six weeks have now elapsed, and I am able to sit out once more in the sunshine. Just opposite me is the steep hillside, grey with shaly rock, and yonder on its flank is the dark cleft which marks the opening of the Blue John Gap. But it is no longer a source of terror. Never again through that ill-omened tunnel shall any strange shape flit out into the world of men. The educated and the scientific, the Dr. Johnsons and the like, may smile at my narrative, but the poorer folk of the countryside had never a doubt as to its truth. On the day after my recovering consciousness they assembled in their hundreds round the Blue John Gap. As the Castleton Courier said:

It was useless for our correspondent, or for any of the adventurous gentlemen who had come from Matlock, Buxton, and other parts, to offer to descend, to explore the cave to the end, and to finally test the extraordinary narrative of Dr. James Hardcastle. The country people had taken the matter into their own hands, and from an early hour of the morning they had worked hard in stopping up the entrance of the tunnel. There is a sharp slope where the shaft begins, and great boulders, rolled along by many willing hands, were thrust down it until the Gap was absolutely sealed. So ends the episode which has caused such excitement throughout the country. Local opinion is fiercely divided upon the subject. On the one hand are those who point to Dr. Hardcastle's impaired health, and to the possibility of cerebral lesions of tubercular origin giving rise to strange hallucinations. Some idee fixe, according to these gentlemen, caused the doctor to wander down the tunnel, and a fall among the rocks was sufficient to account for his injuries. On the other hand, a legend of a strange creature in the Gap has existed for some months back, and the farmers look upon Dr. Hardcastle's narrative and his personal injuries as a final corroboration. So the matter stands, and so the matter will continue to stand, for no definite solution seems to us to be now possible. It transcends human wit to give any scientific explanation which could cover the alleged facts.

Perhaps before the Courier published these words they would have been wise to send their representative to me. I have thought the matter out, as no one else has occasion to do, and it is possible that I might have removed some of the more obvious difficulties of the narrative and brought it one degree nearer to scientific acceptance. Let me then write down the only explanation which seems to me to elucidate what I know to my cost to have been a series of facts. My theory may seem to be wildly improbable, but at least no one can venture to say that it is impossible.

My view is—and it was formed, as is shown by my diary, before my personal adventure—that in this part of England there is a vast subterranean lake or sea, which is fed by the great number of streams which pass down through the limestone. Where there is a large collection of water there must also be some evaporation, mists or rain, and a possibility of vegetation. This in turn suggests that there may be animal life, arising, as the vegetable life would also do, from those seeds and types which had been introduced at an early period of the world's history, when communication with the outer air was more easy. This place had then developed a fauna and flora of its own, including such monsters as the one which I had seen, which may well have been the old cave-bear, enormously enlarged and modified by its new environment. For countless aeons the internal and the external creation had kept apart, growing steadily away from each other. Then there had come some rift in the depths of the mountain which had enabled one creature to wander up and, by means of the Roman tunnel, to reach the open air. Like all subterranean life, it had lost the power of sight, but this had no doubt been compensated for by nature in other directions. Certainly it had some means of finding its way about, and of hunting down the sheep upon the hillside. As to its choice of dark nights, it is part of my theory that light was painful to those great white eyeballs, and that it was only a pitch-black world which it could tolerate. Perhaps, indeed, it was the glare of my lantern which saved my life at that awful moment when we were face to face. So I read the riddle. I leave these facts behind me, and if you can explain them, do so; or if you choose to doubt them, do so. Neither your belief nor your incredulity can alter them, nor affect one whose task is nearly over.

So ended the strange narrative of Dr. James Hardcastle.

Arthur Conan Doyle (1859–1930) created one of the most successful fictional characters of all time: Sherlock Holmes. The immense popularity of Holmes became something of an irritant to his author, since Doyle was keen to write tales about something other than the cases of Baker Street's master detective. Some of Doyle's non-Holmes output was science fiction: his novel *The Lost World* (1912) featuring Professor Challenger—a scientist as irascible as Holmes was coolly rational—became a classic of the genre.

Commentary

Doyle's 1910 story is grounded in reality. The limestone hills around Castleton really are as picturesque as he describes. There really is a semi-precious, purple–blue mineral called Blue John and, in the UK, there are only two places it can be found: Blue John Cavern and Treak Cliff Cavern, both of which hide under a triangular-shaped hill called Treak Cliff a mile or so away from Castleton. And Blue John really was mined. Indeed, extraction continues to this day albeit on a reduced scale—mainly to service the day-trippers who come to explore the show caves and marvel at the Blue John formations. (These popular visitor attractions repeat one of the details mentioned in Doyle's story, namely that the Romans discovered these banded fluorite veins. In truth, there is no evidence the Romans were aware that Blue John existed in Derbyshire. The mineral was probably first mined in about 1750.) The background aspects of "The Terror of Blue John Gap" are, therefore, realistic. But how convincing is the main thrust of the story? Could the caves under Treak Cliff—or caves anywhere else for that matter—be home to a monster?

The monster in Blue John Gap would now be classified as a *cryptid*—an animal, according to the *Oxford English Dictionary*, "whose existence or survival to the present day is disputed or unsubstantiated". Although not part of the *OED* definition, size and aggression surely also play a role: when we talk about unfamiliar life forms we tend to think in terms of *big* creatures that might hurt us. As Doyle demonstrated, they make for the best stories. And there are many such stories. One book devoted to *cryptozoology*—literally, the study of hidden animals—lists more than a thousand cryptids. Some are of world renown: surely everyone has heard of Bigfoot, the Kraken (see Fig. 1.1), and the Loch Ness Monster. Most are of purely local fame—anyone heard of Amaypathenya, Batutut, or Chipekwe? (No, me neither.) What the book demonstrates, however, is that cultures all across the world tell tales of strange lake monsters or fearsome, hairy, manlike beasts. Given the prevalence of these stories, could they be hiding a grain of truth? Are cryptids creeping about out there?

At first glance it seems unlikely.

The ubiquity of monster stories signifies little in itself—after all, the pervasiveness of ghost stories is seldom taken as proof of an afterlife. And, if cryptids *did* exist, is it credible that hard evidence of their existence would be lacking? Earth is now home to 7.5 billion human beings, many of whom possess a smartphone, so if cryptids were around YouTube would be filled with amateur footage of Nessie and the Abominable Snowman.

Fig. 1.1 The legendary Kraken, seizing an unfortunate ship. The artist Edgar Etherington produced this wood engraving to illustrate John Gibson's 1887 book "Monsters of the Sea: Legendary and Authentic" (Credit: Public domain)

And yet …

Biologists accept that they haven't found every large creature currently living on Earth. As I was re-reading Doyle's story, for example, a journal article announced the discovery on the island of Vangunu of a hitherto unknown giant rat—a rodent that lives in trees and feeds on coconuts. Biologists were only searching for the creature because of tales told by natives of the Solomon Islands. Of course, a nut-munching rat isn't as exciting as Doyle's monster but inevitably science will continue to uncover creatures that previously have been hidden from view. Indeed, since the turn of the century dozens of new mammals have been described. That's just land-based mammals. The deep sea offers a vast unexplored reservoir of ecological niches. Scientists have yet to catalogue many of the creatures that live far below the surface, a region where sunlight doesn't penetrate and pressures are high. It's not impossible that some of the creatures who dwell in the deep are large beasts—sea monsters, if you will. (There are countless tales of strange sea creatures. As a young man, serving in the navy, my father kept a diary. An entry from 1960 describes a couple of sailors fishing one day in the Med. One of the sailors caught something and began to pull. A giant tentacle came out of the water, grabbed hold of the fishing line, and snapped it … The shocked anglers dropped their poles and ran.)

Although it would be unwise to bet on the existence of cryptids, and certainly on cryptids of the kind described by Arthur Conan Doyle, it's not entirely unreasonable to believe "monsters" exist somewhere on our planet. And if a creature such the Yeti *were* discovered, just imagine the peaks of excitement that journalists would scale! But what would be the impact of such a discovery on science? I don't believe it would necessarily be profound. Ultimately—depending upon the details, of course—the discovery might merely indicate that creatures can survive in niches away from the ever-increasing influence of humans. Indeed, this thought raises a question. More than a century after the publication of Doyle's story, are there any discoveries in biology that *would* have a profound impact? The answer is: of course! Let's briefly consider one example of just such a profound discovery, based on work that took place in the 1970s. We can then consider a suggestion which, if it were proved true, would represent a biological discovery of far more significance than that of a bipedal carnivore in Blue John Gap.

*   *   *

Big creatures are made up of eukaryotic cells—cells that possess a nucleus, a cytoskeleton, and flexible cell walls (or no cell walls at all). These cells have the ability to combine, pass on genetic information through sex, and evolve different body shapes in response to environmental pressures. The eukaryotic grade of life thus contains all the complex life we see around us, in all its wonderful variety—from aardvarks, bats, and camels through to xerus, yaks, and zebras. Bigfoot, Nessie, and the monster of Blue John Gap would all be examples of the eukaryotic grade of life. By many measures, however, the most successful organisms on Earth aren't the big, complex, life forms. The most successful creatures are single-celled microorganisms like bacteria.

Prokaryotic cells—those that possess rigid cell walls—have been on the planet for 3.6 billion years or so. These cells evolve biochemical rather than morphological responses to environmental pressures, an approach that helped them survive the various mass extinctions which have wiped out so many biologically complex species in the past. Bacteria constitute a large fraction of Earth's biomass. So when we talk about the dominant form of life on Earth we shouldn't be so human-centric: bacteria might be unable to ponder the mysteries of the universe, but they are important. And although land-dwellers with the size of the Blue John Gap monster are unlikely to have eluded us, surprises surely await biologists in the microcosmic world. There's more space for discovery in the world of microorganisms than there is in the ocean depths.

The microbial world has already sprung one big surprise: the revelation that life is so adaptable it can evolve to thrive in niches once thought deadly. Some organisms require high acid levels in order to survive and some require high alkaline levels. Some organisms require the absence rather than the presence of oxygen. Some organisms thrive in salt, or at high temperatures, or in ice. So-called *extremophiles* can be found in deserts, subsurface Antarctic ice, hydrothermal vents … life is pretty much everywhere you look.

In the late 1970s, the American microbiologist Carl Richard Woese discovered that a certain class of extremophile wasn't a type of eukaryote, but neither was it a type of bacteria. This particular prokaryote represented a completely new domain of life. The discovery was as much a shock to the scientific community as the sighting of the Blue John monster was to James Hardcastle. Biologists had got used to the idea of there being two domains of life: eukarya and bacteria. Woese showed there are *three* domains: eukarya, bacteria, and archaea. Archaea and bacteria are no more related to each other than they are to eukaryotes. (Fig. 1.2 shows a simplified version of the tree of life.)

For a while biologists believed all archaea were extremophiles, eking out a living in harsh environments. In recent years, however, archaea have been found in a wide range of moderate habitats; they have even been found in the human gut and on human skin. They are among the most abundant organisms on the planet.

Four decades on and Woese's discovery has transformed our understanding of microbial diversity, provided us with new perspectives on our thinking about evolution, and improved our knowledge of the history of life on Earth. It was a key breakthrough in biology. Is it possible that an even more exciting discovery might be waiting to be made?

*   *   *

Although there are three separate domains—archaea, bacteria, and eukarya—all forms of life on Earth possess certain common features. For instance, all life forms are able to metabolize; in other words, they can draw nutrients from their environment, convert them into energy, and excrete the waste products. And all life forms are able to reproduce. Indeed, one can argue that metabolism and reproduction are necessary factors in order to define something as being alive. But some common features of life on Earth seemingly *aren't* necessary. Some features of life appear to be mere accidents. Consider, for example, chirality.

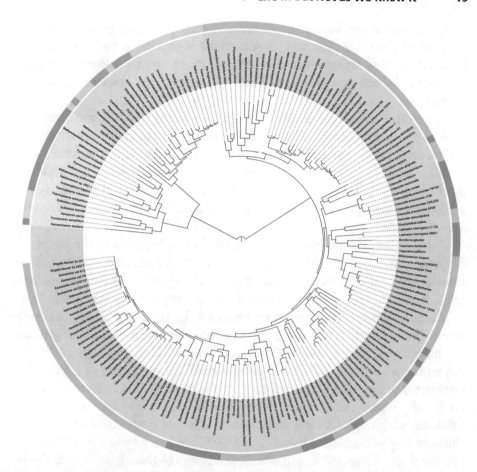

Fig. 1.2   This tree of life shows the relationships between those species with sequenced genomes as of 2006 (the same image now would contain much more detail). The diagram's central point represents LUCA—the last universal common ancestor, the organism from which all current life on Earth ultimately evolved. The three colours represent the three domains of life. Green = archaea (the relatively recent discovery made by Woese). Lilac = bacteria. Pink = eukarya (animals, plants, fungi). The reader with eagle eyes—or a magnifying glass—might make out Homo sapiens in the tree: we are in the pink segment, second from the edge with the lilac. Tree diagrams such as this emphasise just how inconsequential we are in evolutionary terms; we are just one of leaf in a vast forest of species, and each leaf can trace its ancestry back to LUCA (Credit: image generated using the online phylogenetic tree viewer iTOL Interactive Tree of Life; retraced by Mariana Ruiz Villarreal)

Chirality is a fancy word for handedness. If an object is different to its mirror image then it displays chirality; human hands provide the clearest example, since the mirror image of a person's right hand cannot be superimposed on the left hand. Well, the laws of chemistry are blind to chirality. Although many molecules can have a "handedness"—they can come in left-handed or right-handed varieties, just as a glove can be left-handed or right-handed—as far as chemistry is concerned the handedness doesn't much matter. A molecule is a molecule, just as a glove is a glove. In *biochemistry*, however, chirality is crucial.

Large biochemical molecules come in two mirror-image forms, and the molecules must be compatible in order to assemble more complex structures. All life on Earth uses amino acids to construct proteins, and those amino acids are all left-handed. Sugars are all right-handed. And DNA is a right-handed double helix. If life were starting now, from fresh, there would presumably be a 50–50 chance of it using right-handed amino acids or a left-handed double helix for DNA.

Chirality isn't the only example of chance when it comes to the history of life on Earth. Consider the genetic code—a set of rules that enables cells to translate information stored within nucleic acids (DNA or RNA) into proteins. In DNA, information is preserved in the sequence of nucleotide bases —adenine, cytosine, guanine, and thymine; A, C, G, and T. The genetic code is based on nucleotide triplets—called *codons*—and different codons spell out the names of different amino acids. The codon GGU, for example, codes for the amino acid glycine. Within a gene, a particular sequence of triplets dictates the sequence of amino acids that are chained together to form a particular protein—and proteins are the molecules that allow cells to function. (When genetic engineering is mentioned in films or news reports it's often to the backdrop of a string of bases: … ATGGCTAC … or whatever. Whenever you see these strings, it helps to picture groupings of three bases—in the same way that the seemingly meaningless string "thefatredhatwastoobigfortheboy" can be decomposed into meaningful three-letter groupings: "the fat red hat was too big for the boy".)

Apart from some minor exceptions, all life uses the same 20 amino acids to build proteins. The point to make here, however, is that a three-letter genetic code based on four bases could produce up to $4 \times 4 \times 4 = 64$ different amino acids. Life employs only a limited number of amino acids even though further types exist—chemists can synthesise them—and the genetic code is large enough to contain them. Perhaps the particular genetic code used by life is an accident rather than an inevitability.

Or consider the basic chemicals upon which all life on Earth is based: carbon, hydrogen, oxygen, nitrogen, and phosphorus. It's possible to imagine life making use of arsenic instead of phosphorus. (Arsenic is a poison precisely because it acts in a similar fashion to phosphorus. For an arsenic-based life form it would be phosphorus that was the poison.) Or, to take the classic science-fictional scenario (see Chapter 4: "Life on Mars"), perhaps life could make use of silicon rather than carbon: the silicon atom, like carbon, has four electrons in its outermost shell and so could form the rings and long chains required of biological molecules.

And the point of all this? Well, scientists don't understand the details of *abiogenesis*—the original formation of life from inorganic material—but we *do* know that Earth gave rise to life at least once. Eventually this primitive organism, whatever it was, evolved into the so-called last universal common ancestor, which then split into the domains of bacteria and archaea; some time later the more complex eukaryotic cell came into being. Suppose, though, that abiogenesis happened more than once! It's possible. After all, several biologists argue that if environmental conditions on a young planet are such that life *can* come into being then life *will* come into being. This viewpoint, if correct, means it's entirely possible life started from non-life on several different occasions in Earth's distant history. Those ancient organisms would be examples of life—but not life as we know it. Alien life, in a sense. Such life forms might be arsenic based, or use a different genetic code, or require left-handed sugars. Furthermore, those different forms of life would not be in direct competition with familiar life forms—they might not be able to consume the same food, for example and there would be no possibility of swapping genes with everyday organisms. Therefore, "alien" microorganisms descended from other abiogenesis events could in principle still be around today. And because biologists haven't spent much resource in looking for such microorganisms— the deep sea is better explored than this putative "shadow biosphere"—alien life might be widespread. It could even be living on you, right now!

This is, of course, pure speculation. But if it turned out to be true? Well, the discovery of microscopic "alien" organisms descended from different creation events here on Earth—life, but not life as we know it—that would be of tremendous scientific importance. For one thing, it would confirm that the creation of life from non-life can occur in a number of different ways; it would increase the likelihood of there being aliens out there in the Galaxy. And encountering extraterrestrials would be much more exciting than finding a Bigfoot-like creature in a Derbyshire cave.

* * *

Nature perhaps contains only the creatures with which we are by now familiar. But biotechnology is advancing to the stage where scientists can *make* new life forms. Perhaps some future biotechnologist could *create* a creature similar to the monster of the Blue John Gap? We continue that thought in Chapter 2.

Notes and Further Reading

One book devoted to cryptozoology—Eberhart (2002) gives a comprehensive guide to fabled creatures from around the world.

a journal article announced the discovery on the island of Vangunu of a hitherto unknown giant, tree-dwelling rat—See Lavery and Judge (2017) for details of the discovery; Young (2017) nicely puts this discovery in context.

a vast unexplored reservoir of ecological niches—See Palumbi and Palumbi (2015) for an interesting account of the extremes of sea life, including the possibility of unknown species lurking in the depths.

a completely new domain of life—The foundational paper, which outlines the case for a three-domain picture of life, is Woese and Fox (1977). For more information on biodiversity, the different characteristics of different types of organism, and the evolution of life on Earth, see the Tree of Life Web Project (ToL n.d.). This collaborative effort by the worldwide biological community has succeeded in constructing more than 10,000 web pages.

this putative "shadow biosphere"—The term "shadow biosphere" was coined by Cleland and Copley (2005). Carol Cleland, who researches into the philosophical bases of science, has noted (Cleland 2007) that arguments concerning the status of desert varnish—a coating found on exposed rock surfaces in arid environments—have continued since Darwin's time, with some scientists claiming it is a living organism. If we can't be sure whether desert varnish is a life form then surely it's not unreasonable to imagine the possibility of a shadow biosphere? For a review of the possibilities for strange life forms, aimed at the general reader, see Toomey (2014). Considering that this appears to be such a science fictional concept, I'm surprised no well-known SF stories explore the idea (at least, none exist to my knowledge).

Bibliography

Cleland, C.E.: Epistemological issues in the study of microbial life: alternative biospheres. Stud. Hist. Phil. Biol. Biomed. Sci. **38**, 847–861 (2007)

Cleland, C.E., Copley, S.D.: The possibility of alternative microbial life on Earth. Int. J. Astrobiol. **4**(4), 165–173 (2005)

Doyle, A.C.: The terror of Blue John Gap. Strand Magazine. August issue (1910)

Eberhard, G.M.: Mysterious Creatures: A Guide to Cryptozoology. ABC-CLIO, Santa Barbara (2002)

Lavery, T.H., Judge, H.: A new species of giant rat (Muridae, Uromys) from Vangunu, Solomon Islands. J. Mammal. **98**(6), 1518–1530 (2017)

Palumbi, S.R., Palumbi, A.R.: The Extreme Life of the Sea. Princeton University Press, Princeton, NJ (2015)

ToL. Tree of Life Web Project. http://tolweb.org/tree/ (n.d.)

Toomey, D.: Weird Life: The Search for Life That Is Very, Very Different from Our Own. W. W. Norton, New York (2014)

Woese, C.R., Fox, G.E.: Phylogenetic structure of the prokaryotic domain; the primary kingdoms. Proc. Natl. Acad. Sci. USA. **74**(11), 5088–5090 (1977)

Young, E.: Giant, tree-dwelling rat discovered in Solomon Islands. Nature. (2017). https://doi.org/10.1038/nature.2017.22684

2

Transmogrification

The Voice in the Night (William Hope Hodgson)

It was a dark, starless night. We were becalmed in the Northern Pacific. Our exact position I do not know; for the sun had been hidden during the course of a weary, breathless week, by a thin haze which had seemed to float above us, about the height of our mastheads, at whiles descending and shrouding the surrounding sea.

With there being no wind, we had steadied the tiller, and I was the only man on deck. The crew, consisting of two men and a boy, were sleeping forrard in their den; while Will—my friend, and the master of our little craft—was aft in his bunk on the port side of the little cabin.

Suddenly, from out of the surrounding darkness, there came a hail: "Schooner, ahoy!"

The cry was so unexpected that I gave no immediate answer, because of my surprise.

It came again—a voice curiously throaty and inhuman, calling from somewhere upon the dark sea away on our port broadside:

"Schooner, ahoy!"

"Hullo!" I sung out, having gathered my wits somewhat. "What are you? What do you want?"

"You need not be afraid," answered the queer voice, having probably noticed some trace of confusion in my tone. "I am only an old man."

© Springer Nature Switzerland AG 2019
S. Webb, *New Light Through Old Windows: Exploring Contemporary Science Through 12 Classic Science Fiction Tales*, Science and Fiction,
https://doi.org/10.1007/978-3-030-03195-4_2

The pause sounded oddly; but it was only afterwards that it came back to me with any significance.

"Why don't you come alongside, then?" I queried somewhat snappishly; for I liked not his hinting at my having been a trifle shaken.

"I—I—can't. It wouldn't be safe. I—" The voice broke off, and there was silence.

"What do you mean?" I asked, growing more and more astonished. "Why not safe? Where are you?"

I listened for a moment; but there came no answer. And then, a sudden indefinite suspicion, of I knew not what, coming to me, I stepped swiftly to the binnacle, and took out the lighted lamp. At the same time, I knocked on the deck with my heel to waken Will. Then I was back at the side, throwing the yellow funnel of light out into the silent immensity beyond our rail. As I did so, I heard a slight, muffled cry, and then the sound of a splash as though someone had dipped oars abruptly. Yet I cannot say that I saw anything with certainty; save, it seemed to me, that with the first flash of the light, there had been something upon the waters, where now there was nothing.

"Hullo, there!" I called. "What foolery is this!"

But there came only the indistinct sounds of a boat being pulled away into the night.

Then I heard Will's voice, from the direction of the after scuttle:

"What's up, George?"

"Come here, Will!" I said.

"What is it?" he asked, coming across the deck.

I told him the queer thing which had happened. He put several questions; then, after a moment's silence, he raised his hands to his lips, and hailed:

"Boat, ahoy!"

From a long distance away there came back to us a faint reply, and my companion repeated his call. Presently, after a short period of silence, there grew on our hearing the muffled sound of oars; at which Will hailed again.

This time there was a reply:

"Put away the light."

"I'm damned if I will," I muttered; but Will told me to do as the voice bade, and I shoved it down under the bulwarks.

"Come nearer," he said, and the oar-strokes continued. Then, when apparently some half-dozen fathoms distant, they again ceased.

"Come alongside," exclaimed Will. "There's nothing to be frightened of aboard here!"

"Promise that you will not show the light?"

"What's to do with you," I burst out, "that you're so infernally afraid of the light?"

"Because—" began the voice, and stopped short.

"Because what?" I asked quickly.

Will put his hand on my shoulder.

"Shut up a minute, old man," he said, in a low voice. "Let me tackle him."

He leant more over the rail.

"See here, Mister," he said, "this is a pretty queer business, you coming upon us like this, right out in the middle of the blessed Pacific. How are we to know what sort of a hanky-panky trick you're up to? You say there's only one of you. How are we to know, unless we get a squint at you—eh? What's your objection to the light, anyway?"

As he finished, I heard the noise of the oars again, and then the voice came; but now from a greater distance, and sounding extremely hopeless and pathetic.

"I am sorry—sorry! I would not have troubled you, only I am hungry, and—so is she."

The voice died away, and the sound of the oars, dipping irregularly, was borne to us.

"Stop!" sung out Will. "I don't want to drive you away. Come back! We'll keep the light hidden, if you don't like it."

He turned to me:

"It's a damned queer rig, this; but I think there's nothing to be afraid of?"

There was a question in his tone, and I replied.

"No, I think the poor devil's been wrecked around here, and gone crazy." The sound of the oars drew nearer.

"Shove that lamp back in the binnacle," said Will; then he leaned over the rail and listened. I replaced the lamp, and came back to his side. The dipping of the oars ceased some dozen yards distant.

"Won't you come alongside now?" asked Will in an even voice. "I have had the lamp put back in the binnacle."

"I—I cannot," replied the voice. "I dare not come nearer. I dare not even pay you for the—the provisions."

"That's all right," said Will, and hesitated. "You're welcome to as much grub as you can take—" Again he hesitated.

"You are very good," exclaimed the voice. "May God, Who understands everything, reward you—" It broke off huskily.

"The—the lady?" said Will abruptly. "Is she—"

"I have left her behind upon the island," came the voice. "What island?" I cut in.

"I know not its name," returned the voice. "I would to God — —!" it began, and checked itself as suddenly.

"Could we not send a boat for her?" asked Will at this point.

"No!" said the voice, with extraordinary emphasis. "My God! No!" There was a moment's pause; then it added, in a tone which seemed a merited reproach:

"It was because of our want I ventured—because her agony tortured me."

"I am a forgetful brute," exclaimed Will. "Just wait a minute, whoever you are, and I will bring you up something at once."

In a couple of minutes he was back again, and his arms were full of various edibles. He paused at the rail.

"Can't you come alongside for them?" he asked.

"No—I DARE NOT," replied the voice, and it seemed to me that in its tones I detected a note of stifled craving—as though the owner hushed a mortal desire. It came to me then in a flash, that the poor old creature out there in the darkness, was SUFFERING for actual need of that which Will held in his arms; and yet, because of some unintelligible dread, refraining from dashing to the side of our little schooner, and receiving it. And with the lightning-like conviction, there came the knowledge that the Invisible was not mad; but sanely facing some intolerable horror.

"Damn it, Will!" I said, full of many feelings, over which predominated a vast sympathy. "Get a box. We must float off the stuff to him in it."

This we did—propelling it away from the vessel, out into the darkness, by means of a boathook. In a minute, a slight cry from the Invisible came to us, and we knew that he had secured the box.

A little later, he called out a farewell to us, and so heartful a blessing, that I am sure we were the better for it. Then, without more ado, we heard the ply of oars across the darkness.

"Pretty soon off," remarked Will, with perhaps just a little sense of injury.

"Wait," I replied. "I think somehow he'll come back. He must have been badly needing that food."

"And the lady," said Will. For a moment he was silent; then he continued: "It's the queerest thing ever I've tumbled across, since I've been fishing."

"Yes," I said, and fell to pondering.

And so the time slipped away—an hour, another, and still Will stayed with me; for the queer adventure had knocked all desire for sleep out of him.

The third hour was three parts through, when we heard again the sound of oars across the silent ocean.

"Listen!" said Will, a low note of excitement in his voice. "He's coming, just as I thought," I muttered.

The dipping of the oars grew nearer, and I noted that the strokes were firmer and longer. The food had been needed.

They came to a stop a little distance off the broadside, and the queer voice came again to us through the darkness:

"Schooner, ahoy!"

"That you?" asked Will.

"Yes," replied the voice. "I left you suddenly; but—but there was great need."

"The lady?" questioned Will.

"The—lady is grateful now on earth. She will be more grateful soon in—in heaven."

Will began to make some reply, in a puzzled voice; but became confused, and broke off short. I said nothing. I was wondering at the curious pauses, and, apart from my wonder, I was full of a great sympathy.

The voice continued:

"We—she and I, have talked, as we shared the result of God's tenderness and yours—"

Will interposed; but without coherence.

"I beg of you not to—to belittle your deed of Christian charity this night," said the voice. "Be sure that it has not escaped His notice."

It stopped, and there was a full minute's silence. Then it came again:

"We have spoken together upon that which—which has befallen us. We had thought to go out, without telling any, of the terror which has come into our—lives. She is with me in believing that to-night's happenings are under a special ruling, and that it is God's wish that we should tell to you all that we have suffered since—since—"

"Yes?" said Will softly.

"Since the sinking of the Albatross."

"Ah!" I exclaimed involuntarily. "She left Newcastle for 'Frisco some six months ago, and hasn't been heard of since."

"Yes," answered the voice. "But some few degrees to the North of the line she was caught in a terrible storm, and dismasted. When the day came, it was found that she was leaking badly, and, presently, it falling to a calm, the sailors took to the boats, leaving—leaving a young lady—my fiancée—and myself upon the wreck.

"We were below, gathering together a few of our belongings, when they left. They were entirely callous, through fear, and when we came up upon the deck, we saw them only as small shapes afar off upon the horizon. Yet we did not despair, but set to work and constructed a small raft. Upon this we put such few matters as it would hold including a quantity of water and some ship's biscuit. Then, the vessel being very deep in the water, we got ourselves on to the raft, and pushed off.

"It was later, when I observed that we seemed to be in the way of some tide or current, which bore us from the ship at an angle; so that in the course of three hours, by my watch, her hull became invisible to our sight, her broken masts remaining in view for a somewhat longer period. Then, towards evening, it grew misty, and so through the night. The next day we were still encompassed by the mist, the weather remaining quiet.

"For four days we drifted through this strange haze, until, on the evening of the fourth day, there grew upon our ears the murmur of breakers at a distance. Gradually it became plainer, and, somewhat after midnight, it appeared to sound upon either hand at no very great space. The raft was raised upon a swell several times, and then we were in smooth water, and the noise of the breakers was behind.

"When the morning came, we found that we were in a sort of great lagoon; but of this we noticed little at the time; for close before us, through the enshrouding mist, loomed the hull of a large sailing-vessel. With one accord, we fell upon our knees and thanked God; for we thought that here was an end to our perils. We had much to learn.

"The raft drew near to the ship, and we shouted on them to take us aboard; but none answered. Presently the raft touched against the side of the vessel, and, seeing a rope hanging downwards, I seized it and began to climb. Yet I had much ado to make my way up, because of a kind of grey, lichenous fungus which had seized upon the rope, and which blotched the side of the ship lividly.

"I reached the rail and clambered over it, on to the deck. Here I saw that the decks were covered, in great patches, with grey masses, some of them rising into nodules several feet in height; but at the time I thought less of this matter than of the possibility of there being people aboard the ship. I shouted; but none answered. Then I went to the door below the poop deck. I opened it, and peered in. There was a great smell of staleness, so that I knew in a moment that nothing living was within, and with the knowledge, I shut the door quickly; for I felt suddenly lonely.

"I went back to the side where I had scrambled up. My—my sweetheart was still sitting quietly upon the raft. Seeing me look down she called up to know whether there were any aboard of the ship. I replied that the vessel had the appearance of having been long deserted; but that if she would wait a little I would see whether there was anything in the shape of a ladder by which she could ascend to the deck. Then we would make a search through the vessel together. A little later, on the opposite side of the decks, I found a rope side-ladder. This I carried across, and a minute afterwards she was beside me.

"Together we explored the cabins and apartments in the after part of the ship; but nowhere was there any sign of life. Here and there within the cabins themselves, we came across odd patches of that queer fungus; but this, as my sweetheart said, could be cleansed away.

"In the end, having assured ourselves that the after portion of the vessel was empty, we picked our ways to the bows, between the ugly grey nodules of that strange growth; and here we made a further search which told us that there was indeed none aboard but ourselves.

"This being now beyond any doubt, we returned to the stern of the ship and proceeded to make ourselves as comfortable as possible. Together we cleared out and cleaned two of the cabins: and after that I made examination whether there was anything eatable in the ship. This I soon found was so, and thanked God in my heart for His goodness. In addition to this I discovered the whereabouts of the fresh-water pump, and having fixed it I found the water drinkable, though somewhat unpleasant to the taste.

"For several days we stayed aboard the ship, without attempting to get to the shore. We were busily engaged in making the place habitable. Yet even thus early we became aware that our lot was even less to be desired than might have been imagined; for though, as a first step, we scraped away the odd patches of growth that studded the floors and walls of the cabins and saloon, yet they returned almost to their original size within the space of twenty-four hours, which not only discouraged us, but gave us a feeling of vague unease.

"Still we would not admit ourselves beaten, so set to work afresh, and not only scraped away the fungus, but soaked the places where it had been, with carbolic, a can-full of which I had found in the pantry. Yet, by the end of the week the growth had returned in full strength, and, in addition, it had spread to other places, as though our touching it had allowed germs from it to travel elsewhere.

"On the seventh morning, my sweetheart woke to find a small patch of it growing on her pillow, close to her face. At that, she came to me, so soon as she could get her garments upon her. I was in the galley at the time lighting the fire for breakfast.

"Come here, John," she said, and led me aft. When I saw the thing upon her pillow I shuddered, and then and there we agreed to go right out of the ship and see whether we could not fare to make ourselves more comfortable ashore.

"Hurriedly we gathered together our few belongings, and even among these I found that the fungus had been at work; for one of her shawls had a little lump of it growing near one edge. I threw the whole thing over the side, without saying anything to her.

"The raft was still alongside, but it was too clumsy to guide, and I lowered down a small boat that hung across the stern, and in this we made our way to the shore. Yet, as we drew near to it, I became gradually aware that here the vile fungus, which had driven us from the ship, was growing riot. In places it rose into horrible, fantastic mounds, which seemed almost to quiver, as with a quiet life, when the wind blew across them. Here and there it took on the forms of vast fingers, and in others it just spread out flat and smooth and treacherous. Odd places, it appeared as grotesque stunted trees, seeming extraordinarily kinked and gnarled—the whole quaking vilely at times.

"At first, it seemed to us that there was no single portion of the surrounding shore which was not hidden beneath the masses of the hideous lichen; yet, in this, I found we were mistaken; for somewhat later, coasting along the shore at a little distance, we descried a smooth white patch of what appeared to be fine sand, and there we landed. It was not sand. What it was I do not know. All that I have observed is that upon it the fungus will not grow; while every-where else, save where the sand-like earth wanders oddly, path-wise, amid the grey desolation of the lichen, there is nothing but that loathsome greyness.

"It is difficult to make you understand how cheered we were to find one place that was absolutely free from the growth, and here we deposited our belongings. Then we went back to the ship for such things as it seemed to us we should need. Among other matters, I managed to bring ashore with me one of the ship's sails, with which I constructed two small tents, which, though exceedingly rough-shaped, served the purpose for which they were intended. In these we lived and stored our various necessities, and thus for a matter of some four weeks all went smoothly and without particular unhappiness. Indeed, I may say with much of happiness—for—for we were together.

"It was on the thumb of her right hand that the growth first showed. It was only a small circular spot, much like a little grey mole. My God! how the fear leapt to my heart when she showed me the place. We cleansed it, between us, washing it with carbolic and water. In the morning of the following day she showed her hand to me again. The grey warty thing had returned. For a little while, we looked at one another in silence. Then, still wordless, we started again to remove it. In the midst of the operation she spoke suddenly.

"'What's that on the side of your face, dear?' Her voice was sharp with anxi-ety. I put my hand up to feel.

"'There! Under the hair by your ear. A little to the front a bit.' My finger rested upon the place, and then I knew.

"'Let us get your thumb done first,' I said. And she submitted, only because she was afraid to touch me until it was cleansed. I finished washing and disin-fecting her thumb, and then she turned to my face. After it was finished we sat together and talked awhile of many things for there had come into our lives sudden, very terrible thoughts. We were, all at once, afraid of something worse than death. We spoke of loading the boat with provisions and water and making our way out on to the sea; yet we were helpless, for many causes, and—and the growth had attacked us already. We decided to stay. God would do with us what was His will. We would wait.

"A month, two months, three months passed and the places grew somewhat, and there had come others. Yet we fought so strenuously with the fear that its headway was but slow, comparatively speaking.

"Occasionally we ventured off to the ship for such stores as we needed. There we found that the fungus grew persistently. One of the nodules on the maindeck became soon as high as my head.

"We had now given up all thought or hope of leaving the island. We had realized that it would be unallowable to go among healthy humans, with the things from which we were suffering.

"With this determination and knowledge in our minds we knew that we should have to husband our food and water; for we did not know, at that time, but that we should possibly live for many years.

"This reminds me that I have told you that I am an old man. Judged by the years this is not so. But—but—"

He broke off; then continued somewhat abruptly:

"As I was saying, we knew that we should have to use care in the matter of food. But we had no idea then how little food there was left of which to take care. It was a week later that I made the discovery that all the other bread tanks—which I had supposed full—were empty, and that (beyond odd tins of vegetables and meat, and some other matters) we had nothing on which to depend, but the bread in the tank which I had already opened.

"After learning this I bestirred myself to do what I could, and set to work at fishing in the lagoon; but with no success. At this I was somewhat inclined to feel desperate until the thought came to me to try outside the lagoon, in the open sea.

"Here, at times, I caught odd fish; but so infrequently that they proved of but little help in keeping us from the hunger which threatened.

It seemed to me that our deaths were likely to come by hunger, and not by the growth of the thing which had seized upon our bodies.

"We were in this state of mind when the fourth month wore out. When I made a very horrible discovery. One morning, a little before midday. I came off from the ship with a portion of the biscuits which were left. In the mouth of her tent I saw my sweetheart sitting, eating something.

"'What is it, my dear?' I called out as I leapt ashore. Yet, on hearing my voice, she seemed confused, and, turning, slyly threw something towards the edge of the little clearing. It fell short, and a vague suspicion having arisen within me, I walked across and picked it up. It was a piece of the grey fungus.

"As I went to her with it in my hand, she turned deadly pale; then rose red.

"I felt strangely dazed and frightened.

"'My dear! My dear!' I said, and could say no more. Yet at words she broke down and cried bitterly. Gradually, as she calmed, I got from her the news that she had tried it the preceding day, and—and liked it. I got her to promise on her knees not to touch it again, however great our hunger. After she had promised she told me that the desire for it had come suddenly, and that, until the moment of desire, she had experienced nothing towards it but the most extreme repulsion.

"Later in the day, feeling strangely restless, and much shaken with the thing which I had discovered, I made my way along one of the twisted paths— formed by the white, sand-like substance—which led among the fungoid growth. I had, once before, ventured along there; but not to any great distance. This time, being involved in perplexing thought, I went much further than hitherto.

"Suddenly I was called to myself by a queer hoarse sound on my left. Turning quickly I saw that there was movement among an extraordinarily shaped mass of fungus, close to my elbow. It was swaying uneasily, as though it possessed life of its own. Abruptly, as I stared, the thought came to me that the thing had a grotesque resemblance to the figure of a distorted human creature. Even as the fancy flashed into my brain, there was a slight, sickening noise of tearing, and I saw that one of the branch-like arms was detaching itself from the surrounding grey masses, and coming towards me. The head of the thing—a shapeless grey ball, inclined in my direction. I stood stupidly, and the vile arm brushed across my face. I gave out a frightened cry, and ran back a few paces. There was a sweetish taste upon my lips where the thing had touched me. I licked them, and was immediately filled with an inhuman desire. I turned and seized a mass of the fungus. Then more and—more. I was insatiable. In the midst of devouring, the remembrance of the morning's discovery swept into my mazed brain. It was sent by God. I dashed the fragment I held to the ground. Then, utterly wretched and feeling a dreadful guiltiness, I made my way back to the little encampment.

"I think she knew, by some marvellous intuition which love must have given, so soon as she set eyes on me. Her quiet sympathy made it easier for me, and I told her of my sudden weakness; yet omitted to mention the extraordinary thing which had gone before. I desired to spare her all unnecessary terror.

"But, for myself, I had added an intolerable knowledge, to breed an incessant terror in my brain; for I doubted not but that I had seen the end of one of those men who had come to the island in the ship in the lagoon; and in that monstrous ending I had seen our own.

"Thereafter we kept from the abominable food, though the desire for it had entered into our blood. Yet our drear punishment was upon us; for, day by day, with monstrous rapidity, the fungoid growth took hold of our poor bodies. Nothing we could do would check it materially, and so—and so—we who had been human, became— Well, it matters less each day. Only—only we had been man and maid!

"And day by day the fight is more dreadful, to withstand the hungerlust for the terrible lichen.

"A week ago we ate the last of the biscuit, and since that time I have caught three fish. I was out here fishing tonight when your schooner drifted upon me out of the mist. I hailed you. You know the rest, and may God, out of His great heart, bless you for your goodness to a—a couple of poor outcast souls." There was the dip of an oar—another. Then the voice came again, and for the last time, sounding through the slight surrounding mist, ghostly and mournful.

"God bless you! Good-bye!"

"Good-bye," we shouted together, hoarsely, our hearts full of many emotions.

I glanced about me. I became aware that the dawn was upon us.

The sun flung a stray beam across the hidden sea; pierced the mist dully, and lit up the receding boat with a gloomy fire. Indistinctly I saw something nodding between the oars. I thought of a sponge—a great, grey nodding sponge—The oars continued to ply. They were grey—as was the boat—and my eyes searched a moment vainly for the conjunction of hand and oar. My gaze flashed back to the—head. It nodded forward as the oars went backward for the stroke. Then the oars were dipped, the boat shot out of the patch of light, and the—the thing went nodding into the mist.

William Hope Hodgson (1877–1918) was a prolific author whose fiction, which often featured a nautical setting, tended to be concerned with the weird and the fantastic. Several modern SF and fantasy authors cite Hodgson as an influence, and his writing is said to have inspired H.P. Lovecraft. Hodgson was killed by an artillery shell at Ypres, during World War I.

Commentary

William Hope Hodgson had a particular interest in the transformation of the human body. From childhood, Hodgson was fascinated by the sea (stay with me—the connection with bodily transformation will become clear) and when he was just 13 he ran away to join the navy. He was caught and forcibly returned home, but a few months later his father allowed him to begin an apprenticeship as a cabin boy. Hodgson, however, was short in stature—even as an adult he was only 5′ 4″—and he was, furthermore, in possession of movie-star good looks. This combination almost guaranteed that his fellow crew members would bully him. And they did. Most men might have given up and jumped ship. Hodgson, though, took up bodybuilding. It wasn't long before any seaman trying to bully Hodgson found to his cost he was dealing with, pound-for-pound, the strongest man in the navy. Hodgson went on to found a "School of Physical Culture", where he trained others—including police officers—to strengthen their bodies and improve their physique.

Hodgson's method of implementing physical change through exercise depended of course upon conscious control. The horrifying bodily changes described in his 1907 story "The Voice in the Night" were involuntary—the result of a terrible fungal infection. Although I'm not aware of there being anything quite like the fungus described in this story, there is at least one disfiguring disease that causes similar visual outcomes. Indeed, I wonder whether an encounter with a sufferer during one of Hodgson's voyages might have sown in his imagination the seed of the story?

The disease epidermodysplasia verruciformis (EV)—also referred to as Lewandowsky–Lutz dysplasia (after the dermatologists who first described it in 1922) and as treeman syndrome (after the visual appearance of sufferers)—is an extremely rare disorder. The medical literature contains only a few hundred reported cases. Currently, perhaps the most notable case is that of Abul Bajandar from Bangladesh (see Fig. 2.1). However, the best documented case of EV is that of Dede Koswara, an Indonesian man whose arms and feet were covered in the warts and thick horns that typify the disease. Koswara died from complications of EV in 2016, aged just 42. During his relatively short life he was the subject of numerous television documentaries, and no viewer could consider his plight to be anything other than both heartbreaking and shocking. Much of the skin on Koswara's body had the appearance of bark; his hands and feet resembled the gnarled branches of old trees. When I look at photographs of Koswara, I'm put in mind of the unfortunate couple in Hodgson's story.

Fig. 2.1 A photograph of Abul Bajandar taken in February 2016, when Bajandar was 25 years old. The lesions on his hands are clearly visible. In 2017, surgeons at Dhaka Medical College succeeded in removing 5 kg of the wart-like growths during 16 operations on Bajandar. Unfortunately, by 2018 the growths had started to reappear (Credit: Monirul Alam)

The terrible skin lesions characteristic of EV are caused by repeated infection with the human papillomavirus—a virus with many subtypes, some of which can persist and cause warts. If a person has an abnormal or impaired immune response to certain subtypes of the wart virus then, in some unlucky cases, the result can be EV. In some individuals with EV the impaired immune response is the result of HIV infection or of drugs used to suppress rejection following organ transplants. The disease is then said to be acquired. In most patients, however, the disease is the result of an inherited condition. The problem in these cases is caused ultimately by a mutation in certain genes, with the so-called *EVER1* or *EVER2* genes being particularly implicated. The condition is recessive, which means that in order to be affected a person must receive two copies of the abnormal gene on a non-sex chromosome; if only one abnormal copy is inherited then the individual becomes a carrier rather than a sufferer. So a child born to two carriers has a 25% chance of being completely unaffected, a 50% chance of becoming a carrier, and a 25% chance of being affected. Dede Koswara was one of the unlucky individuals who received two copies of an abnormal gene, one each from carrier parents.

Treatment of the condition is about alleviating symptoms. Surgeons can sometimes operate to remove the lesions, and this might permit a patient to recover the use of fingers and feet, but even if surgery is successful there can be no guarantee the warts won't return. There is no cure for EV.

Bioscience, however, progresses at a dizzying pace. The same morning I sat down to write this commentary to Hodgson's creepy story, news was published of an exciting development in gene editing techniques. The development represents the first step on a journey that might lead not only to a cure for diseases such as EV but also to more general control over the workings—and even appearance—of the human body.

A team of Chinese bioscientists made the advance. They were investigating the rare but serious blood disorder β-thalassaemia, which is caused by a so-called point mutation: it requires the base adenine (A) to be switched to the base guanine (G) at a particular point in the *HBB* gene. This mutation does occur in the Chinese population, but it's relatively rare; it's much rarer to have someone receive two copies of the mutation—and, as with EV, the disease is recessive so a person must receive two copies of the mutated gene for the disorder to manifest itself. The Chinese team wanted to investigate the disorder in embryos, so their first challenge was simply to source embryos that had two copies of this rare mutation. The team proceeded by identifying a patient with β-thalassaemia, extracting some of the patient's skin cells, and then developing 20 embryonic clones each with two copies of the mutation. They then employed a new base editing technique, a procedure that has been called "chemical surgery". The team built on a gene-editing method called CRISPR-Cas9, which was developed in 2012 and allows bioscientists to use enzymes to cut genes with accuracy. The Chinese team's modification of the CRISPR-Cas9 technique can guide an enzyme to specific gene sequences, and then simply swap G to A (or the base cytosine (C) to thymine (T)). No cutting involved. In 8 out of the 20 cloned embryos, the offending G could be switched to A at the relevant spot in the HBB gene. And fixing just one of the two faulty genes would be enough to cure a recessive disease.

Clearly, a huge amount of work must take place before this technique can be used in medicine; many questions are still to be answered. But this proof-of-principle experiment shows how base-editing techniques might correct a disease mutation. In future, rare diseases such as treeman syndrome might be prevented by editing an at-risk embryo; if the disease is already present in an adult, techniques might be developed to swap out the faulty genes. If the unfortunate married couple in Hodgson's story were living today, they might hope for a cure.

But, as with every scientific advance, there are dangers.

* * *

For centuries people have been using artificial devices to transform the human body, and we can expect this phenomenon to continue. Some transformations are entirely unremarkable (vision correction, for example); some are currently noteworthy (advanced prosthetics); some are seemingly science-fictional (the addition of non-anthropomorphic structures such as wings). We can expect some future humans to undergo a level of "cyborgisation" because throughout history people have used whatever technology has been available in order to alter their bodies. Ultimately, though, these transmogrifications are transformations of the individual. What makes the advances in genetic engineering so powerful—and potentially so dangerous—is that they have the capacity to affect the entire human species.

Science fiction writers have been in the vanguard of examining the opportunities and threats of biotechnology as applied to human transmogrification. They've contemplated changes in form even more bizarre than those imagined by Hodgson in "The Voice in the Night". As long ago as 1983 Greg Bear, in his award-winning novelette "Blood Music" (later expanded into a novel of the same name), discussed one possibility.

In Bear's story a rebellious bioengineer manufactures a simple biological computer based upon his own white blood cells. His employer, nervous about the direction of this work, orders him to destroy the computers; instead, the bioengineer injects the computers into his bloodstream and smuggles them out of the company lab so he can continue work on them in his own time. Once inside his body, however, these artificial cells evolve and develop self-awareness. They start to alter their own genetic material, and the bioengineer begins to transmogrify. At first, it's all good: better eyesight, greater strength, improved skin, quicker brain. He infects others and they find that *their* health improves, their muscles strengthen, their intelligence increases—again, all good. Except that these mutant cells infect more people, then more, and eventually go on to assimilate much of the biosphere!

Bear's story was one of the first visions of the *grey-goo scenario*—the notion that technology at the molecular or nanoscale level (the realm where biotechnology takes place) might cause the end of the world through a runaway effect.

The threat of a gray-goo doomsday, and similar runaway effects, can surely be alleviated. If society pays serious attention to the problem *now*, while we have the luxury of time, then safeguards can be put in place—just as engineers implement safeguards for all new technologies. The power of biotechnology, however, gives rise to one problem humanity might struggle to avoid.

The problem involves the quite staggering rate of progress in biotechnology. To appreciate the scale of this progress, recall that the double-helix structure of DNA was discovered as recently as 1953. The Human Genome Project—the campaign to determine the sequence of base pairs that make up human DNA—was launched in 1990, cost about $2.7 billion, and took more than a decade to succeed. Today, companies can sequence your genome within about an hour for $1000 or so. These unprecedented advances in biotechnology have occurred within a typical person's lifespan: 1953 was the year of Queen Elizabeth's coronation, after all, and as I write Elizabeth is still on her throne. If this rate of progress continues then the cost of sequencing a genome will soon be pennies.

Here is another example of progress in biotechnology. Starting in 1989, bioscientists began to investigate gene function by using "gene targeting"— they'd inactivate, or "knock out", an entire gene in an animal, replace it with a piece of artificial DNA, and then observe any effects on the animal. In 2012, this expensive and time-consuming technique was revolutionised with the introduction of the CRISPR-Cas9 gene editing technique mentioned above. In 2017, as we've seen, scientists showed how to swap individual base pairs. In the future, it seems likely that any reasonably competent undergraduate biology student will be able to tinker with the fundamental chemicals that define the make-up of a human being.

So what's the problem? These biotechnological advances will, after all, transform medicine: people will have access to personalised treatments, dreadful diseases such as EV will be eradicated, and it will be difficult for readers to even imagine the sort of world described in "The Voice in the Night". Unfortunately, those same biotechnological advances will become so cheap and so widespread they'll be readily available to terrorists, criminals, or simply the unhinged. The bad guys as well as the good will have ability to alter existing life or to create new life. Unfortunately it's not difficult to imagine a future fanatic creating a deadly pathogen in the kitchen of his bedsit, and releasing it on an unsuspecting population. How can we protect ourselves against such a threat? We should start thinking about it now.

*   *   *

If an artificial pathogen *were* inflicted upon us, by either accident or design, what would happen to our civilisation? Well, we might face a world not dissimilar to the one depicted in Chapter 3.

Notes and Further Reading

the transformation of the human body—Bodily transformation is a venerable theme in literature. The Latin poet Ovid wrote beautifully about such flux two thousand years ago in *Metamorphoses*, which ranks as one of the most influential books ever written. Melville (2008) is a good English translation of *Metamorphoses*. Marina Warner (2004), in a wide-ranging volume, analyses themes of transformation in literature since the time of Ovid.

In *Shapeshifters* (2018) Gavin Francis, a doctor, has written about human alteration from the perspective of a medical professional. His patients—pregnant women; menopausal women; those suffering from anorexia; people transitioning—visit his clinic in the hope he can influence what is changing in their bodies.

Hodgson was fascinated by the sea—Sam Gifford (n.d.) runs a blog that contains a large amount of interesting information about William Hope Hodgson.

the best documented case of EV is that of Dede Koswara—The unfortunate Koswara was the subject of a hard-to-watch documentary called *Half Man, Half Tree* (TLC 2007) and soon after another documentary called *Tree Man* (ABC 2008).

a procedure that has been called "chemical surgery"—About two-thirds of documented human diseases are the result of the sort of point mutations described in the text. So base editing, if it becomes routine, could give rise to exciting developments in medicine. The correction of β-thalassaemia in embryos using a base editing technique is described in Liang et al. (2017).

as with every scientific advance, there are dangers—There are many potential abuses of biotechnology. One obvious cause of concern, which I haven't had the space to discuss in the main text, is the prospect of "designer babies". Science fiction writers have been interested in this since at least the 1930s, when Aldous Huxley (1932) published his satirical novel *Brave New World*. The development of CRISPR-Cas9, and the subsequent development of base-swapping, means there will come a time when embryos can be edited. Whether this prospect is years or decades away depends on advances in technology (these techniques are, after all, still in their infancy) and on the regulatory framework imposed by governments (obvious ethical concerns surrounding gene editing are combined with a general public distrust). But it seems likely that at some point, in a lab under some jurisdiction somewhere in the world, scientists will be allowed to manipulate the basic building blocks of human life in order to prevent or treat genetic disease. The worry then becomes that embryos will eventually be edited not only to remove the possibility of disease but to possess features society deems attractive—blue eyes, blond hair, improved athletic ability, or those features that happen to be in fashion. Several ethical issues spring to mind with this scenario. Will the health gap between rich and poor increase within society and between nations? Will parents be influenced by society to choose various features for their children? If it turns out, as seems likely, that there's a complex relationship between genes and traits such as personality and intelligence, will people start "tweaking" embryos just to try and give their offspring an advantage in the genetic sweepstake—and what if that tweaking goes wrong? And then there are the unforeseeable but probably inevitable unintended consequences of dabbling with DNA …

Even if the technology behind gene editing turns out to be ineffective, society will still need to debate the potential promise and peril of designer babies because a quite different genetic advance provides an efficient and cost-effective method to allow parents (or the directors of hatcheries, as in *Brave New World*) to choose traits for their babies. Instead of getting parents to produce a single embryo, and then editing the genetic information, the simpler approach is for them to produce lots of embryos and screen all of them for their genetic potential. The desirability of an embryo marked as high-IQ, low-risk for asthma, high-risk for male-pattern baldness might be weighed against an embryo possessing the possibility of athletic prowess but low-IQ and a susceptibility to diabetes. Choices, choices. Society must soon debate these ethical and moral issues in earnest.

For centuries people have been using artificial devices to transform the human body—The use of spectacles or contact lenses to improve vision is of course entirely unremarkable; the use of prosthetic arms and legs to improve the quality of life for those who have lost limbs is becoming more common; soon we shall see the everyday application of devices to augment the natural abilities of individuals who are entirely healthy. It seems likely that in the future many people will choose to undergo some form of "cyborgisation". Professor Hugh Herr, head of the biomechatronics group at the MIT Media Lab, researches this idea. Herr himself uses specialised prostheses, the result of having both legs amputated below the knee following a climbing accident. (The story of Herr's accident, and his single-minded struggle to recover from the loss of his legs, is recounted in Osius (1991).) Using these devices Herr can effectively change his height, and attempt climbs that would previously have been impossible. In a sense, Herr has taken a first step towards cyborgisation. At TED2018 (available online; see Herr (2018)) he talked about his vision for humanity in the 21st century: "Humans may … extend their bodies into non-anthropomorphic structures, such as wings, controlling and feeling each wing movement within the nervous system. … During the twilight years of this century, I believe humans will be unrecognizable in morphology and dynamics from what we are today." This idea, of course, has been investigated by SF writers. In *Man Plus*, for example, Frederik Pohl (1976) described how an astronaut's body might be transformed for the rigours of life on the Martian surface: genitals would be removed, cloaca added, lungs revamped … Advances in biotechnology mean our species might son change in ways we can barely foresee.

Greg Bear, in his award-winning novelette—See Bear (1983).

Bibliography

ABC: *Tree Man*. TV documentary. Episode of Medical Mysteries. ABC, New York (2008)

Bear, G.: Blood music. Analog Science Fact & Fiction. June issue (1983)

Francis, G.: Shapeshifters: On Medicine & Human Change. Wellcome, London (2018)

Gifford, S.: William Hope Hodgson. Website. https://williamhopehodgson.wordpress.com (n. d.)

Herr, H.: How we'll become cyborgs and extend human potential. TED 2018. https://www.ted.com/talks/hugh_herr_how_we_ll_become_cyborgs_and_extend_human_potential/ (2018)

Hodgson, W.H.: The voice in the night. Blue Book Magazine. November issue (1907)

Huxley, A.: Brave New World. Chatto and Windus, London (1932)

Liang, P., Ding, C., Sun, H., et al.: Correction of β-thalassaemia mutant by base editor in human embryo. Protein Cell. **8**, 811–822 (2017)

Melville, A.D. (transl): Ovid. Metamorphoses. OUP, Oxford (2008)

Osius, A.: Second Ascent. The Story of Hugh Herr. Stackpole Books, Mechanicsburg, PA (1991)

Pohl, F.: Man Plus. Random House, New York (1976)

TLC: Half Man, Half Tree. TV documentary. Episode of My Shocking Story. The Learning Channel, Silver Spring, MA (2007)

Warner, M.: Fantastic Metamorphoses, Other Worlds: Ways of Telling the Self. OUP, Oxford (2004)

3

Pandemic

The Scarlet Plague (Jack London)

I

THE way led along upon what had once been the embankment of a railroad. But no train had run upon it for many years. The forest on either side swelled up the slopes of the embankment and crested across it in a green wave of trees and bushes. The trail was as narrow as a man's body, and was no more than a wild-animal runway. Occasionally, a piece of rusty iron, showing through the forest-mould, advertised that the rail and the ties still remained. In one place, a ten-inch tree, bursting through at a connection, had lifted the end of a rail clearly into view. The tie had evidently followed the rail, held to it by the spike long enough for its bed to be filled with gravel and rotten leaves, so that now the crumbling, rotten timber thrust itself up at a curious slant. Old as the road was, it was manifest that it had been of the monorail type.

An old man and a boy travelled along this runway. They moved slowly, for the old man was very old, a touch of palsy made his movements tremulous, and he leaned heavily upon his staff. A rude skullcap of goat-skin protected his head from the sun. From beneath this fell a scant fringe of stained and dirty-white hair. A visor, ingeniously made from a large leaf, shielded his eyes, and from under this he peered at the way of his feet on the trail. His beard, which should have been snow-white but which showed the same weather-wear

© Springer Nature Switzerland AG 2019
S. Webb, *New Light Through Old Windows: Exploring Contemporary Science Through 12 Classic Science Fiction Tales*, Science and Fiction,
https://doi.org/10.1007/978-3-030-03195-4_3

and camp-stain as his hair, fell nearly to his waist in a great tangled mass. About his chest and shoulders hung a single, mangy garment of goat-skin. His arms and legs, withered and skinny, betokened extreme age, as well as did their sunburn and scars and scratches betoken long years of exposure to the elements.

The boy, who led the way, checking the eagerness of his muscles to the slow progress of the elder, likewise wore a single garment—a ragged-edged piece of bear-skin, with a hole in the middle through which he had thrust his head. He could not have been more than twelve years old. Tucked coquettishly over one ear was the freshly severed tail of a pig. In one hand he carried a medium-sized bow and an arrow.

On his back was a quiverful of arrows. From a sheath hanging about his neck on a thong, projected the battered handle of a hunting knife. He was as brown as a berry, and walked softly, with almost a catlike tread. In marked contrast with his sunburned skin were his eyes—blue, deep blue, but keen and sharp as a pair of gimlets. They seemed to bore into aft about him in a way that was habitual. As he went along he smelled things, as well, his distended, quivering nostrils carrying to his brain an endless series of messages from the outside world. Also, his hearing was acute, and had been so trained that it operated automatically. Without conscious effort, he heard all the slight sounds in the apparent quiet—heard, and differentiated, and classified these sounds—whether they were of the wind rustling the leaves, of the humming of bees and gnats, of the distant rumble of the sea that drifted to him only in lulls, or of the gopher, just under his foot, shoving a pouchful of earth into the entrance of his hole.

Suddenly he became alertly tense. Sound, sight, and odor had given him a simultaneous warning. His hand went back to the old man, touching him, and the pair stood still. Ahead, at one side of the top of the embankment, arose a crackling sound, and the boy's gaze was fixed on the tops of the agitated bushes. Then a large bear, a grizzly, crashed into view, and likewise stopped abruptly, at sight of the humans. He did not like them, and growled querulously. Slowly the boy fitted the arrow to the bow, and slowly he pulled the bowstring taut. But he never removed his eyes from the bear.

The old man peered from under his green leaf at the danger, and stood as quietly as the boy. For a few seconds this mutual scrutinizing went on; then, the bear betraying a growing irritability, the boy, with a movement of his head, indicated that the old man must step aside from the trail and go down the embankment. The boy followed, going backward, still holding the bow taut and ready. They waited till a crashing among the bushes from the opposite side of the embankment told them the bear had gone on. The boy grinned as he led back to the trail.

"A big un, Granser," he chuckled.

The old man shook his head.

"They get thicker every day," he complained in a thin, undependable falsetto. "Who'd have thought I'd live to see the time when a man would be afraid of his life on the way to the Cliff House. When I was a boy, Edwin, men and women and little babies used to come out here from San Francisco by tens of thousands on a nice day. And there weren't any bears then. No, sir. They used to pay money to look at them in cages, they were that rare."

"What is money, Granser?"

Before the old man could answer, the boy recollected and triumphantly shoved his hand into a pouch under his bear-skin and pulled forth a battered and tarnished silver dollar. The old man's eyes glistened, as he held the coin close to them.

"I can't see," he muttered. "You look and see if you can make out the date, Edwin."

The boy laughed.

"You're a great Granser," he cried delightedly, "always making believe them little marks mean something."

The old man manifested an accustomed chagrin as he brought the coin back again close to his own eyes.

"2012," he shrilled, and then fell to cackling grotesquely. "That was the year Morgan the Fifth was appointed President of the United States by the Board of Magnates. It must have been one of the last coins minted, for the Scarlet Death came in 2013. Lord! Lord!—think of it! Sixty years ago, and I am the only person alive today that lived in those times. Where did you find it, Edwin?"

The boy, who had been regarding him with the tolerant curiousness one accords to the prattlings of the feeble-minded, answered promptly.

"I got it off of Hoo-Hoo. He found it when we was herdin' goats down near San José last spring. Hoo-Hoo said it was *money*. Ain't you hungry, Granser?"

The ancient caught his staff in a tighter grip and urged along the trail, his old eyes shining greedily.

"I hope Hare-Lip's found a crab... or two," he mumbled, "They're good eating, crabs, mighty good eating when you've no more teeth and you've got grandsons that love their old grandsire and make a point of catching crabs for him. When I was a boy—"

But Edwin, suddenly stopped by what he saw, was drawing the bowstring on a fitted arrow. He had paused on the brink of a crevasse in the embankment.

An ancient culvert had here washed out, and the stream, no longer confined, had cut a passage through the fill. On the opposite side, the end of a rail projected and overhung. It showed rustily through the creeping vines which overran it. Beyond, crouching by a bush, a rabbit looked across at him in trembling hesitancy. Fully fifty feet was the distance, but the arrow flashed true; and the transfixed rabbit, crying out in sudden fright and hurt, struggled painfully away into the brush. The boy himself was a flash of brown skin and flying fur as he bounded down the steep wall of the gap and up the other side. His lean muscles were springs of steel that released into graceful and efficient action. A hundred feet beyond, in a tangle of bushes, he overtook the wounded creature, knocked its head on a convenient tree-trunk, and turned it over to Granser to carry.

"Rabbit is good, very good," the ancient quavered, "but when it comes to a toothsome delicacy I prefer crab. When I was a boy—"

"Why do you say so much that ain't got no sense?" Edwin impatiently interrupted the other's threatened garrulousness.

The boy did not exactly utter these words, but something that remotely resembled them and that was more guttural and explosive and economical of qualifying phrases. His speech showed distant kinship with that of the old man, and the latter's speech was approximately an English that had gone through a bath of corrupt usage.

"What I want to know," Edwin continued, "is why you call crab 'toothsome delicacy'? Crab is crab, ain't it? No one I never heard calls it such funny things."

The old man sighed but did not answer, and they moved on in silence. The surf grew suddenly louder, as they emerged from the forest upon a stretch of sand dunes bordering the sea. A few goats were browsing among the sandy hillocks, and a skin-clad boy, aided by a wolfish-looking dog that was only faintly reminiscent of a collie, was watching them. Mingled with the roar of the surf was a continuous, deep-throated barking or bellowing, which came from a cluster of jagged rocks a hundred yards out from shore. Here huge sea-lions hauled themselves up to lie in the sun or battle with one another. In the immediate foreground arose the smoke of a fire, tended by a third savage-looking boy. Crouched near him were several wolfish dogs similar to the one that guarded the goats.

The old man accelerated his pace, sniffing eagerly as he neared the fire. "Mussels!" he muttered ecstatically. "Mussels! And ain't that a crab, Hoo-Hoo? Ain't that a crab? My, my, you boys are good to your old grandsire." Hoo-Hoo, who was apparently of the same age as Edwin, grinned. "All you want, Granser. I got four."

The old man's palsied eagerness was pitiful. Sitting down in the sand as quickly as his stiff limbs would let him, he poked a large rock-mussel from out of the coals. The heat had forced its shells apart, and the meat, salmon-colored, was thoroughly cooked. Between thumb and forefinger, in trembling haste, he caught the morsel and carried it to his mouth. But it was too hot, and the next moment was violently ejected. The old man spluttered with the pain, and tears ran out of his eyes and down his cheeks.

The boys were true savages, possessing only the cruel humor of the savage. To them the incident was excruciatingly funny, and they burst into loud laughter. Hoo-Hoo danced up and down, while Edwin rolled gleefully on the ground. The boy with the goats came running to join in the fun.

"Set 'em to cool, Edwin, set 'em to cool," the old man besought, in the midst of his grief, making no attempt to wipe away the tears that still flowed from his eyes. "And cool a crab, Edwin, too. You know your grandsire likes crabs."

From the coals arose a great sizzling, which proceeded from the many mussels bursting open their shells and exuding their moisture. They were large shellfish, running from three to six inches in length. The boys raked them out with sticks and placed them on a large piece of driftwood to cool.

"When I was a boy, we did not laugh at our elders; we respected them."

The boys took no notice, and Granser continued to babble an incoherent flow of complaint and censure. But this time he was more careful, and did not burn his mouth. All began to eat, using nothing but their hands and making loud mouth-noises and lip-smackings. The third boy, who was called Hare-Lip, slyly deposited a pinch of sand on a mussel the ancient was carrying to his mouth; and when the grit of it bit into the old fellow's mucous membrane and gums, the laughter was again uproarious. He was unaware that a joke had been played on him, and spluttered and spat until Edwin, relenting, gave him a gourd of fresh water with which to wash out his mouth.

"Where's them crabs, Hoo-Hoo?" Edwin demanded. "Granser's set upon having a snack."

Again Granser's eyes burned with greediness as a large crab was handed to him. It was a shell with legs and all complete, but the meat had long since departed. With shaky fingers and babblings of anticipation, the old man broke off a leg and found it filled with emptiness.

"The crabs, Hoo-Hoo?" he wailed. "The crabs?"

"I was fooling Granser. They ain't no crabs! I never found one."

The boys were overwhelmed with delight at sight of the tears of senile disappointment that dribbled down the old man's cheeks. Then, unnoticed, Hoo-Hoo replaced the empty shell with a fresh-cooked crab. Already dismembered, from the cracked legs the white meat sent forth a small cloud of savory steam. This attracted the old man's nostrils, and he looked down in amazement.

The change of his mood to one of joy was immediate. He snuffled and muttered and mumbled, making almost a croon of delight, as he began to eat. Of this the boys took little notice, for it was an accustomed spectacle. Nor did they notice his occasional exclamations and utterances of phrases which meant nothing to them, as, for instance, when he smacked his lips and champed his gums while muttering: "Mayonnaise! Just think—mayonnaise! And it's sixty years since the last was ever made! Two generations and never a smell of it! Why, in those days it was served in every restaurant with crab." When he could eat no more, the old man sighed, wiped his hands on his naked legs, and gazed out over the sea. With the content of a full stomach, he waxed reminiscent.

"To think of it! I've seen this beach alive with men, women, and children on a pleasant Sunday. And there weren't any bears to eat them up, either. And right up there on the cliff was a big restaurant where you could get anything you wanted to eat. Four million people lived in San Francisco then. And now, in the whole city and county there aren't forty all told. And out there on the sea were ships and ships always to be seen, going in for the Golden Gate or coming out. And airships in the air—dirigibles and flying machines. They could travel two hundred miles an hour. The mail contracts with the New York and San Francisco Limited demanded that for the minimum. There was a chap, a Frenchman, I forget his name, who succeeded in making three hundred; but the thing was risky, too risky for conservative persons. But he was on the right clew, and he would have managed it if it hadn't been for the Great Plague. When I was a boy, there were men alive who remembered the coming of the first aeroplanes, and now I have lived to see the last of them, and that sixty years ago."

The old man babbled on, unheeded by the boys, who were long accustomed to his garrulousness, and whose vocabularies, besides, lacked the greater portion of the words he used. It was noticeable that in these rambling soliloquies his English seemed to recrudesce into better construction and phraseology. But when he talked directly with the boys it lapsed, largely, into their own uncouth and simpler forms.

"But there weren't many crabs in those days," the old man wandered on. "They were fished out, and they were great delicacies. The open season was only a month long, too. And now crabs are accessible the whole year around. Think of it—catching all the crabs you want, any time you want, in the surf of the Cliff House beach!"

A sudden commotion among the goats brought the boys to their feet. The dogs about the fire rushed to join their snarling fellow who guarded the goats, while the goats themselves stampeded in the direction of their human protectors. A half dozen forms, lean and gray, glided about on the sand hillocks and faced the bristling dogs. Edwin arched an arrow that fell short. But Hare-Lip, with a sling such as David carried into battle against Goliath, hurled a stone through the air that whistled from the speed of its flight. It fell squarely among the wolves and caused them to slink away toward the dark depths of the eucalyptus forest.

The boys laughed and lay down again in the sand, while Granser sighed ponderously. He had eaten too much, and, with hands clasped on his paunch, the fingers interlaced, he resumed his maunderings.

"The fleeting systems lapse like foam," he mumbled what was evidently a quotation. "That's it—foam, and fleeting. All man's toil upon the planet was just so much foam. He domesticated the serviceable animals, destroyed the hostile ones, and cleared the land of its wild vegetation. And then he passed, and the flood of primordial life rolled back again, sweeping his handiwork away—the weeds and the forest inundated his fields, the beasts of prey swept over his flocks, and now there are wolves on the Cliff House beach." He was appalled by the thought. "Where four million people disported themselves, the wild wolves roam today, and the savage progeny of our loins, with prehistoric weapons, defend themselves against the fanged despoilers. Think of it! And all because of the Scarlet Death—"

The adjective had caught Hare-Lip's ear.

"He's always saying that," he said to Edwin. "What is *scarlet?*"

"The scarlet of the maples can shake me like the cry of bugles going by," the old man quoted.

"It's red," Edwin answered the question. "And you don't know it because you come from the Chauffeur Tribe. They never did know nothing, none of them. Scarlet is red—I know that."

"Red is red, ain't it?" Hare-Lip grumbled. "Then what's the good of gettin' cocky and calling it scarlet?"

"Granser, what for do you always say so much what nobody knows?" he asked. "Scarlet ain't anything, but red is red. Why don't you say red, then?"

"Red is not the right word," was the reply. "The plague was scarlet. The whole face and body turned scarlet in an hour's time. Don't I know? Didn't I see enough of it? And I am telling you it was scarlet because—well, because it *was* scarlet. There is no other word for it."

"Red is good enough for me," Hare-Lip muttered obstinately. "My dad calls red red, and he ought to know. He says everybody died of the Red Death."

"Your dad is a common fellow, descended from a common fellow," Granser retorted heatedly. "Don't I know the beginnings of the Chauffeurs? Your grandsire was a chauffeur, a servant, and without education. He worked for other persons. But your grandmother was of good stock, only the children did not take after her. Don't I remember when I first met them, catching fish at Lake Temescal?"

"What is *education?*" Edwin asked.

"Calling red scarlet," Hare-Lip sneered, then returned to the attack on Granser. "My dad told me, an' he got it from his dad afore he croaked, that your wife was a Santa Rosan, an' that she was sure no account. He said she was a *hash-slinger* before the Red Death, though I don't know what a *hash-slinger* is. You can tell me, Edwin."

But Edwin shook his head in token of ignorance.

"It is true, she was a waitress," Granser acknowledged. "But she was a good woman, and your mother was her daughter. Women were very scarce in the days after the Plague. She was the only wife I could find, even if she was a *hash-slinger*, as your father calls it. But it is not nice to talk about our progenitors that way."

"Dad says that the wife of the first Chauffeur was a *lady*—"

"What's a *lady?*" Hoo-Hoo demanded.

"A *lady's* a Chauffeur squaw," was the quick reply of Hare-Lip.

"The first Chauffeur was Bill, a common fellow, as I said before," the old man expounded; "but his wife was a lady, a great lady. Before the Scarlet Death she was the wife of Van Worden. He was President of the Board of Industrial Magnates, and was one of the dozen men who ruled America. He was worth one billion, eight hundred millions of dollars—coins like you have there in your pouch, Edwin. And then came the Scarlet Death, and his wife became the wife of Bill, the first Chauffeur. He used to beat her, too. I have seen it myself."

Hoo-Hoo, lying on his stomach and idly digging his toes in the sand, cried out and investigated, first, his toe-nail, and next, the small hole he had dug. The other two boys joined him, excavating the sand rapidly with their hands till there lay three skeletons exposed. Two were of adults, the third being that of a part-grown child. The old man hudged along on the ground and peered at the find.

"Plague victims," he announced. "That's the way they died everywhere in the last days. This must have been a family, running away from the contagion and perishing here on the Cliff House beach. They—what are you doing, Edwin?"

This question was asked in sudden dismay, as Edwin, using the back of his hunting knife, began to knock out the teeth from the jaws of one of the skulls.

"Going to string 'em," was the response.

The three boys were now hard at it; and quite a knocking and hammering arose, in which Granser babbled on unnoticed.

"You are true savages. Already has begun the custom of wearing human teeth. In another generation you will be perforating your noses and ears and wearing ornaments of bone and shell. I know. The human race is doomed to sink back farther and farther into the primitive night ere again it begins its bloody climb upward to civilization. When we increase and feel the lack of room, we will proceed to kill one another. And then I suppose you will wear human scalp-locks at your waist, as well—as you, Edwin, who are the gentlest of my grandsons, have already begun with that vile pigtail. Throw it away, Edwin, boy; throw it away."

"What a gabble the old geezer makes," Hare-Lip remarked, when, the teeth all extracted, they began an attempt at equal division.

They were very quick and abrupt in their actions, and their speech, in moments of hot discussion over the allotment of the choicer teeth, was truly a gabble. They spoke in monosyllables and short jerky sentences that was more a gibberish than a language. And yet, through it ran hints of grammatical construction, and appeared vestiges of the conjugation of some superior culture. Even the speech of Granser was so corrupt that were it put down literally it would be almost so much nonsense to the reader. This, however, was when he talked with the boys.

When he got into the full swing of babbling to himself, it slowly purged itself into pure English. The sentences grew longer and were enunciated with a rhythm and ease that was reminiscent of the lecture platform.

"Tell us about the Red Death, Granser," Hare-Lip demanded, when the teeth affair had been satisfactorily concluded.

"The Scarlet Death," Edwin corrected.

"An' don't work all that funny lingo on us," Hare-Lip went on. "Talk sensible, Granser, like a Santa Rosan ought to talk. Other Santa Rosans don't talk like you."

II

THE old man showed pleasure in being thus called upon. He cleared his throat and began.

"Twenty or thirty years ago my story was in great demand. But in these days nobody seems interested—"

"There you go!" Hare-Lip cried hotly. "Cut out the funny stuff and talk sensible. What's *interested*? You talk like a baby that don't know how."

"Let him alone," Edwin urged, "or he'll get mad and won't talk at all. Skip the funny places. We'll catch on to some of what he tells us."

"Let her go, Granser," Hoo-Hoo encouraged; for the old man was already maundering about the disrespect for elders and the reversion to cruelty of all humans that fell from high culture to primitive conditions.

The tale began.

"There were very many people in the world in those days. San Francisco alone held four millions—"

"What is millions?" Edwin interrupted. Granser looked at him kindly.

"I know you cannot count beyond ten, so I will tell you. Hold up your two hands. On both of them you have altogether ten fingers and thumbs. Very well. I now take this grain of sand—you hold it, Hoo-Hoo." He dropped the grain of sand into the lad's palm and went on. "Now that grain of sand stands for the ten fingers of Edwin. I add another grain. That's ten more fingers. And I add another, and another, and another, until I have added as many grains as Edwin has fingers and thumbs. That makes what I call one hundred. Remember that word—one hundred. Now I put this pebble in Hare-Lip's hand. It stands for ten grains of sand, or ten tens of fingers, or one hundred fingers. I put in ten pebbles. They stand for a thousand fingers. I take a mussel-shell, and it stands for ten pebbles, or one hundred grains of sand, or one thousand fingers...." And so on, laboriously, and with much reiteration, he strove to build up in their minds a crude conception of numbers. As the quantities increased, he had the boys holding different magnitudes in each of their hands. For still higher sums, he laid the symbols on the log of driftwood; and for symbols he was hard put, being compelled to use the teeth from the skulls for millions, and the crab-shells for billions. It was here that he stopped, for the boys were showing signs of becoming tired.

"There were four million people in San Francisco—four teeth."

The boys' eyes ranged along from the teeth and from hand to hand, down through the pebbles and sand-grains to Edwin's fingers. And back again they ranged along the ascending series in the effort to grasp such inconceivable numbers.

"That was a lot of folks, Granser," Edwin at last hazarded.

"Like sand on the beach here, like sand on the beach, each grain of sand a man, or woman, or child. Yes, my boy, all those people lived right here in San Francisco. And at one time or another all those people came out on this very beach—more people than there are grains of sand. More—more—more. And San Francisco was a noble city. And across the bay—where we camped last year, even more people lived, clear from Point Richmond, on the level ground and on the hills, all the way around to San Leandro—one great city of seven million people.—Seven teeth... there, that's it, seven millions."

Again the boys' eyes ranged up and down from Edwin's fingers to the teeth on the log.

"The world was full of people. The census of 2010 gave eight billions for the whole world—eight crab-shells, yes, eight billions. It was not like today. Mankind knew a great deal more about getting food. And the more food there was, the more people there were. In the year 1800, there were one hundred and seventy millions in Europe alone. One hundred years later—a grain of sand, Hoo-Hoo—one hundred years later, at 1900, there were five hundred millions in Europe—five grains of sand, Hoo-Hoo, and this one tooth. This shows how easy was the getting of food, and how men increased. And in the year 2000 there were fifteen hundred millions in Europe. And it was the same all over the rest of the world. Eight crab-shells there, yes, eight billion people were alive on the earth when the Scarlet Death began.

"I was a young man when the Plague came—twenty-seven years old; and I lived on the other side of San Francisco Bay, in Berkeley. You remember those great stone houses, Edwin, when we came down the hills from Contra Costa? That was where I lived, in those stone houses. I was a professor of English literature."

Much of this was over the heads of the boys, but they strove to comprehend dimly this tale of the past.

"What was them stone houses for?" Hare-Lip queried.

"You remember when your dad taught you to swim?" The boy nodded. "Well, in the University of California—that is the name we had for the houses—we taught young men and women how to think, just as I have taught you now, by sand and pebbles and shells, to know how many people lived in those days. There was very much to teach. The young men and women we taught were called students. We had large rooms in which we taught. I talked to them, forty or fifty at a time, just as I am talking to you now. I told them about the books other men had written before their time, and even, sometimes, in their time—"

"Was that all you did?—just talk, talk, talk?" Hoo-Hoo demanded. "Who hunted your meat for you? and milked the goats? and caught the fish?"

"A sensible question, Hoo-Hoo, a sensible question. As I have told you, in those days food-getting was easy. We were very wise. A few men got the food for many men. The other men did other things. As you say, I talked. I talked all the time, and for this food was given me—much food, fine food, beautiful food, food that I have not tasted in sixty years and shall never taste again. I sometimes think the most wonderful achievement of our tremendous civiliza-tion was food—its inconceivable abundance, its infinite variety, its marvellous delicacy. O my grandsons, life was life in those days, when we had such won-derful things to eat."

This was beyond the boys, and they let it slip by, words and thoughts, as a mere senile wandering in the narrative.

"Our food-getters were called *freemen*. This was a joke. We of the ruling classes owned all the land, all the machines, everything. These food-getters were our slaves. We took almost all the food they got, and left them a little so that they might eat, and work, and get us more food—"

"I'd have gone into the forest and got food for myself," Hare-Lip announced; "and if any man tried to take it away from me, I'd have killed him."

The old man laughed.

"Did I not tell you that we of the ruling class owned all the land, all the forest, everything? Any food-getter who would not get food for us, him we punished or compelled to starve to death. And very few did that. They pre-ferred to get food for us, and make clothes for us, and prepare and administer to us a thousand—a mussel-shell, Hoo-Hoo—a thousand satisfactions and delights. And I was Professor Smith in those days—Professor James Howard Smith. And my lecture courses were very popular—that is, very many of the young men and women liked to hear me talk about the books other men had written. "And I was very happy, and I had beautiful things to eat. And my hands were soft, because I did no work with them, and my body was clean all over and dressed in the softest garments—"

He surveyed his mangy goat-skin with disgust.

"We did not wear such things in those days. Even the slaves had better gar-ments. And we were most clean. We washed our faces and hands often every day. You boys never wash unless you fall into the water or go swimming."

"Neither do you Granser," Hoo-Hoo retorted.

"I know, I know, I am a filthy old man, but times have changed. Nobody washes these days, there are no conveniences. It is sixty years since I have seen a piece of soap.

"You do not know what soap is, and I shall not tell you, for I am telling the story of the Scarlet Death. You know what sickness is. We called it a disease. Very many of the diseases came from what we called germs. Remember that word—germs. A germ is a very small thing. It is like a woodtick, such as you find on the dogs in the spring of the year when they run in the forest. Only the germ is very small. It is so small that you cannot see it—"

Hoo-Hoo began to laugh.

"You're a queer un, Granser, talking about things you can't see. If you can't see 'em, how do you know they are? That's what I want to know. How do you know anything you can't see?"

"A good question, a very good question, Hoo-Hoo. But we did see—some of them. We had what we called microscopes and ultramicroscopes, and we put them to our eyes and looked through them, so that we saw things larger than they really were, and many things we could not see without the microscopes at all. Our best ultramicroscopes could make a germ look forty thousand times larger. A mussel-shell is a thousand fingers like Edwin's. Take forty mussel-shells, and by as many times larger was the germ when we looked at it through a microscope. And after that, we had other ways, by using what we called moving pictures, of making the forty-thousand-times germ many, many thousand times larger still. And thus we saw all these things which our eyes of themselves could not see. Take a grain of sand. Break it into ten pieces. Take one piece and break it into ten. Break one of those pieces into ten, and one of those into ten, and one of those into ten, and one of those into ten, and do it all day, and maybe, by sunset, you will have a piece as small as one of the germs." The boys were openly incredulous. Hare-Lip sniffed and sneered and Hoo-Hoo snickered, until Edwin nudged them to be silent.

"The woodtick sucks the blood of the dog, but the germ, being so very small, goes right into the blood of the body, and there it has many children. In those days there would be as many as a billion—a crab-shell, please—as many as that crab-shell in one man's body. We called germs micro-organisms. When a few million, or a billion, of them were in a man, in all the blood of a man, he was sick. These germs were a disease. There were many different kinds of them—more different kinds than there are grains of sand on this beach. We knew only a few of the kinds. The micro-organic world was an invisible world, a world we could not see, and we knew very little about it. Yet we did know something. There was the *bacillus anthracis*; there was the *micrococcus*; there was the *Bacterium termo*, and the *Bacterium lactis*—that's what turns the goat milk sour even to this day, Hare-Lip; and there were *Schizomycetes* without end. And there were many others...."

Here the old man launched into a disquisition on germs and their natures, using words and phrases of such extraordinary length and meaninglessness, that the boys grinned at one another and looked out over the deserted ocean till they forgot the old man was babbling on.

"But the Scarlet Death, Granser," Edwin at last suggested.

Granser recollected himself, and with a start tore himself away from the rostrum of the lecture-hall, where, to another world audience, he had been expounding the latest theory, sixty years gone, of germs and germ-diseases.

"Yes, yes, Edwin; I had forgotten. Sometimes the memory of the past is very strong upon me, and I forget that I am a dirty old man, clad in goat-skin, wandering with my savage grandsons who are goatherds in the primeval wilderness. 'The fleeting systems lapse like foam,' and so lapsed our glorious, colossal civilization. I am Granser, a tired old man. I belong to the tribe of Santa Rosans. I married into that tribe. My sons and daughters married into the Chauffeurs, the Sacramentos, and the Palo-Altos. You, Hare-Lip, are of the Chauffeurs. You, Edwin, are of the Sacramentos. And you, Hoo-Hoo, are of the Palo-Altos. Your tribe takes its name from a town that was near the seat of another great institution of learning. It was called Stanford University. Yes, I remember now. It is perfectly clear. I was telling you of the Scarlet Death. Where was I in my story?"

"You was telling about germs, the things you can't see but which make men sick," Edwin prompted.

"Yes, that's where I was. A man did not notice at first when only a few of these germs got into his body. But each germ broke in half and became two germs, and they kept doing this very rapidly so that in a short time there were many millions of them in the body. Then the man was sick. He had a disease, and the disease was named after the kind of a germ that was in him. It might be measles, it might be influenza, it might be yellow fever; it might be any of thousands and thousands of kinds of diseases.

"Now this is the strange thing about these germs. There were always new ones coming to live in men's bodies. Long and long and long ago, when there were only a few men in the world, there were few diseases. But as men increased and lived closely together in great cities and civilizations, new diseases arose, new kinds of germs entered their bodies. Thus were countless millions and billions of human beings killed. And the more thickly men packed together, the more terrible were the new diseases that came to be. Long before my time, in the middle ages, there was the Black Plague that swept across Europe. It swept across Europe many times. There was tuberculosis, that entered into men wherever they were thickly packed. A hundred years before my time there was the bubonic plague. And in Africa was the sleeping sickness. The bacteriologists fought all these sicknesses and destroyed them, just as you boys fight the wolves away from your goats, or squash the mosquitoes that light on you. The bacteriologists—"

"But, Granser, what is a what-you-call-it?" Edwin interrupted.

"You, Edwin, are a goatherd. Your task is to watch the goats. You know a great deal about goats. A bacteriologist watches germs. That's his task, and he knows a great deal about them. So, as I was saying, the bacteriologists fought with the germs and destroyed them—sometimes. There was leprosy, a horrible disease. A hundred years before I was born, the bacteriologists discovered the germ of leprosy. They knew all about it. They made pictures of it. I have seen those pictures. But they never found a way to kill it. But in 1984, there was the Pantoblast Plague, a disease that broke out in a country called Brazil and that killed millions of people. But the bacteriologists found it out, and found the way to kill it, so that the Pantoblast Plague went no farther. They made what they called a serum, which they put into a man's body and which killed the pantoblast germs without killing the man. And in 1910, there was Pellagra, and also the hookworm. These were easily killed by the bacteriologists. But in 1947 there arose a new disease that had never been seen before. It got into the bodies of babies of only ten months old or less, and it made them unable to move their hands and feet, or to eat, or anything; and the bacteriologists were eleven years in discovering how to kill that particular germ and save the babies.

"In spite of all these diseases, and of all the new ones that continued to arise, there were more and more men in the world. This was because it was easy to get food. The easier it was to get food, the more men there were; the more men there were, the more thickly were they packed together on the earth; and the more thickly they were packed, the more new kinds of germs became diseases. There were warnings. Soldervetzsky, as early as 1929, told the bacteriologists that they had no guaranty against some new disease, a thousand times more deadly than any they knew, arising and killing by the hundreds of millions and even by the billion. You see, the micro-organic world remained a mystery to the end. They knew there was such a world, and that from time to time armies of new germs emerged from it to kill men. "And that was all they knew about it. For all they knew, in that invisible micro-organic world there might be as many different kinds of germs as there are grains of sand on this beach. And also, in that same invisible world it might well be that new kinds of germs came to be. It might be there that life originated—the 'abysmal fecundity,' Soldervetzsky called it, applying the words of other men who had written before him...."

It was at this point that Hare-Lip rose to his feet, an expression of huge contempt on his face.

"Granser," he announced, "you make me sick with your gabble. Why don't you tell about the Red Death? If you ain't going to, say so, an' we'll start back for camp."

The old man looked at him and silently began to cry. The weak tears of age rolled down his cheeks and all the feebleness of his eighty-seven years showed in his grief-stricken countenance.

"Sit down," Edwin counselled soothingly. "Granser's all right. He's just gettin' to the Scarlet Death, ain't you, Granser? He's just goin' to tell us about it right now. Sit down, Hare-Lip. Go ahead, Granser."

III

THE old man wiped the tears away on his grimy knuckles and took up the tale in a tremulous, piping voice that soon strengthened as he got the swing of the narrative.

"It was in the summer of 2013 that the Plague came. I was twenty-seven years old, and well do I remember it. Wireless despatches—"

Hare-Lip spat loudly his disgust, and Granser hastened to make amends. "We talked through the air in those days, thousands and thousands of miles. And the word came of a strange disease that had broken out in New York. There were seventeen millions of people living then in that noblest city of America. Nobody thought anything about the news. It was only a small thing. There had been only a few deaths. It seemed, though, that they had died very quickly, and that one of the first signs of the disease was the turning red of the face and all the body. Within twenty-four hours came the report of the first case in Chicago. And on the same day, it was made public that London, the greatest city in the world, next to Chicago, had been secretly fighting the plague for two weeks and censoring the news despatches—that is, not permitting the word to go forth to the rest of the world that London had the plague.

"It looked serious, but we in California, like everywhere else, were not alarmed. We were sure that the bacteriologists would find a way to overcome this new germ, just as they had overcome other germs in the past. But the trouble was the astonishing quickness with which this germ destroyed human beings, and the fact that it inevitably killed any human body it entered. No one ever recovered. There was the old Asiatic cholera, when you might eat dinner with a well man in the evening, and the next morning, if you got up early enough, you would see him being hauled by your window in the death-cart. But this new plague was quicker than that—much quicker.

"From the moment of the first signs of it, a man would be dead in an hour. Some lasted for several hours. Many died within ten or fifteen minutes of the appearance of the first signs.

"The heart began to beat faster and the heat of the body to increase. Then came the scarlet rash, spreading like wildfire over the face and body. Most persons never noticed the increase in heat and heart-beat, and the first they knew was when the scarlet rash came out. Usually, they had convulsions at the time of the appearance of the rash. But these convulsions did not last long and were not very severe. If one lived through them, he became perfectly quiet, and only did he feel a numbness swiftly creeping up his body from the feet. The heels became numb first, then the legs, and hips, and when the numbness reached as high as his heart he died. They did not rave or sleep. Their minds always remained cool and calm up to the moment their heart numbed and stopped. And another strange thing was the rapidity of decomposition. No sooner was a person dead than the body seemed to fall to pieces, to fly apart, to melt away even as you looked at it. That was one of the reasons the plague spread so rapidly. All the billions of germs in a corpse were so immediately released.

"And it was because of all this that the bacteriologists had so little chance in fighting the germs. They were killed in their laboratories even as they studied the germ of the Scarlet Death. They were heroes. As fast as they perished, others stepped forth and took their places. It was in London that they first isolated it. The news was telegraphed everywhere. Trask was the name of the man who succeeded in this, but within thirty hours he was dead. Then came the struggle in all the laboratories to find something that would kill the plague germs. All drugs failed. You see, the problem was to get a drug, or serum, that would kill the germs in the body and not kill the body. They tried to fight it with other germs, to put into the body of a sick man germs that were the enemies of the plague germs—"

"And you can't see these germ-things, Granser," Hare-Lip objected, "and here you gabble, gabble, gabble about them as if they was anything, when they're nothing at all. Anything you can't see, ain't, that's what. Fighting things that ain't with things that ain't! They must have been all fools in them days. That's why they croaked. I ain't goin' to believe in such rot, I tell you that."

Granser promptly began to weep, while Edwin hotly took up his defence. "Look here, Hare-Lip, you believe in lots of things you can't see."

Hare-Lip shook his head.

"You believe in dead men walking about. You never seen one dead man walk about."

"I tell you I seen 'em, last winter, when I was wolf-hunting with dad."

"Well, you always spit when you cross running water," Edwin challenged. "That's to keep off bad luck," was Hare-Lip's defence.

"You believe in bad luck?"

"Sure."

"An' you ain't never seen bad luck," Edwin concluded triumphantly. "You're just as bad as Granser and his germs. You believe in what you don't see. Go on, Granser."

Hare-Lip, crushed by this metaphysical defeat, remained silent, and the old man went on. Often and often, though this narrative must not be clogged by the details, was Granser's tale interrupted while the boys squabbled among themselves. Also, among themselves they kept up a constant, low-voiced exchange of explanation and conjecture, as they strove to follow the old man into his unknown and vanished world.

"The Scarlet Death broke out in San Francisco. The first death came on a Monday morning. By Thursday they were dying like flies in Oakland and San Francisco. They died everywhere—in their beds, at their work, walking along the street. It was on Tuesday that I saw my first death—Miss Collbran, one of my students, sitting right there before my eyes, in my lecture-room. I noticed her face while I was talking. It had suddenly turned scarlet. I ceased speaking and could only look at her, for the first fear of the plague was already on all of us and we knew that it had come. The young women screamed and ran out of the room. So did the young men run out, all but two. Miss Collbran's convulsions were very mild and lasted less than a minute. One of the young men fetched her a glass of water. She drank only a little of it, and cried out:

"'My feet! All sensation has left them.'

"After a minute she said, 'I have no feet. I am unaware that I have any feet. And my knees are cold. I can scarcely feel that I have knees.'

"She lay on the floor, a bundle of notebooks under her head. And we could do nothing. The coldness and the numbness crept up past her hips to her heart, and when it reached her heart she was dead. In fifteen minutes, by the clock—I timed it—she was dead, there, in my own classroom, dead. And she was a very beautiful, strong, healthy young woman. And from the first sign of the plague to her death only fifteen minutes elapsed. That will show you how swift was the Scarlet Death.

"Yet in those few minutes I remained with the dying woman in my classroom, the alarm had spread over the university; and the students, by thousands, all of them, had deserted the lecture-room and laboratories. When I emerged, on my way to make report to the President of the Faculty, I found the university deserted. Across the campus were several stragglers hurrying for their homes. Two of them were running.

"President Hoag, I found in his office, all alone, looking very old and very gray, with a multitude of wrinkles in his face that I had never seen before. At the sight of me, he pulled himself to his feet and tottered away to the inner office, banging the door after him and locking it. You see, he knew I had been exposed, and he was afraid. He shouted to me through the door to go away. I shall never forget my feelings as I walked down the silent corridors and out across that deserted campus. I was not afraid. I had been exposed, and I looked upon myself as already dead. It was not that, but a feeling of awful depression that impressed me. Everything had stopped. It was like the end of the world to me—my world. I had been born within sight and sound of the university. It had been my predestined career. My father had been a professor there before me, and his father before him. For a century and a half had this university, like a splendid machine, been running steadily on. And now, in an instant, it had stopped. It was like seeing the sacred flame die down on some thrice-sacred altar. I was shocked, unutterably shocked.

"When I arrived home, my housekeeper screamed as I entered, and fled away. And when I rang, I found the housemaid had likewise fled. I investigated. In the kitchen I found the cook on the point of departure. But she screamed, too, and in her haste dropped a suitcase of her personal belongings and ran out of the house and across the grounds, still screaming. I can hear her scream to this day. You see, we did not act in this way when ordinary diseases smote us. We were always calm over such things, and sent for the doctors and nurses who knew just what to do. But this was different. It struck so suddenly, and killed so swiftly, and never missed a stroke. When the scarlet rash appeared on a person's face, that person was marked by death. There was never a known case of a recovery.

"I was alone in my big house. As I have told you often before, in those days we could talk with one another over wires or through the air. The telephone bell rang, and I found my brother talking to me. He told me that he was not coming home for fear of catching the plague from me, and that he had taken our two sisters to stop at Professor Bacon's home. He advised me to remain where I was, and wait to find out whether or not I had caught the plague.

"To all of this I agreed, staying in my house and for the first time in my life attempting to cook. And the plague did not come out on me. By means of the telephone I could talk with whomsoever I pleased and get the news. Also, there were the newspapers, and I ordered all of them to be thrown up to my door so that I could know what was happening with the rest of the world.

"New York City and Chicago were in chaos. And what happened with them was happening in all the large cities. A third of the New York police were dead. Their chief was also dead, likewise the mayor. All law and order had ceased. The bodies were lying in the streets un-buried. All railroads and vessels carrying food and such things into the great city had ceased running and mobs of the hungry poor were pillaging the stores and warehouses. Murder and robbery and drunkenness were everywhere. Already the people had fled from the city by millions—at first the rich, in their private motor-cars and dirigibles, and then the great mass of the population, on foot, carrying the plague with them, themselves starving and pillaging the farmers and all the towns and villages on the way.

"The man who sent this news, the wireless operator, was alone with his instrument on the top of a lofty building. The people remaining in the city— he estimated them at several hundred thousand—had gone mad from fear and drink, and on all sides of him great fires were raging. He was a hero, that man who staid by his post—an obscure newspaperman, most likely.

"For twenty-four hours, he said, no transatlantic airships had arrived, and no more messages were coming from England. He did state, though, that a message from Berlin—that's in Germany—announced that Hoffmeyer, a bacteriologist of the Metchnikoff School, had discovered the serum for the plague. That was the last word, to this day, that we of America ever received from Europe. If Hoffmeyer discovered the serum, it was too late, or otherwise, long ere this, explorers from Europe would have come looking for us. We can only conclude that what happened in America happened in Europe, and that, at the best, some several score may have survived the Scarlet Death on that whole continent.

"For one day longer the despatches continued to come from New York. Then they, too, ceased. The man who had sent them, perched in his lofty building, had either died of the plague or been consumed in the great conflagrations he had described as raging around him. And what had occurred in New York had been duplicated in all the other cities. It was the same in San Francisco, and Oakland, and Berkeley. By Thursday the people were dying so rapidly that their corpses could not be handled, and dead bodies lay everywhere. Thursday night the panic outrush for the country began. Imagine, my grandsons, people, thicker than the salmon-run you have seen on the Sacramento river, pouring out of the cities by millions, madly over the country, in vain attempt to escape the ubiquitous death. You see, they carried the germs with them. Even the airships of the rich, fleeing for mountain and desert fastnesses, carried the germs.

"Hundreds of these airships escaped to Hawaii, and not only did they bring the plague with them, but they found the plague already there before them. This we learned, by the despatches, until all order in San Francisco vanished, and there were no operators left at their posts to receive or send. It was amazing, astounding, this loss of communication with the world. It was exactly as if the world had ceased, been blotted out. For sixty years that world has no longer existed for me. I know there must be such places as New York, Europe, Asia, and Africa; but not one word has been heard of them—not in sixty years. With the coming of the Scarlet Death the world fell apart, absolutely, irretrievably. Ten thousand years of culture and civilization passed in the twinkling of an eye, 'lapsed like foam.'

"I was telling about the airships of the rich. They carried the plague with them and no matter where they fled, they died. I never encountered but one survivor of any of them—Mungerson. He was afterwards a Santa Rosan, and he married my eldest daughter. He came into the tribe eight years after the plague. He was then nineteen years old, and he was compelled to wait twelve years more before he could marry. You see, there were no unmarried women, and some of the older daughters of the Santa Rosans were already bespoken. So he was forced to wait until my Mary had grown to sixteen years. It was his son, Gimp-Leg, who was killed last year by the mountain lion.

"Mungerson was eleven years old at the time of the plague. His father was one of the Industrial Magnates, a very wealthy, powerful man. It was on his airship, the Condor, that they were fleeing, with all the family, for the wilds of British Columbia, which is far to the north of here. But there was some accident, and they were wrecked near Mount Shasta. You have heard of that mountain. It is far to the north. The plague broke out amongst them, and this boy of eleven was the only survivor. For eight years he was alone, wandering over a deserted land and looking vainly for his own kind. And at last, travelling south, he picked up with us, the Santa Rosans.

"But I am ahead of my story. When the great exodus from the cities around San Francisco Bay began, and while the telephones were still working, I talked with my brother. I told him this flight from the cities was insanity, that there were no symptoms of the plague in me, and that the thing for us to do was to isolate ourselves and our relatives in some safe place. We decided on the Chemistry Building, at the university, and we planned to lay in a supply of provisions, and by force of arms to prevent any other persons from forcing their presence upon us after we had retired to our refuge.

"All this being arranged, my brother begged me to stay in my own house for at least twenty-four hours more, on the chance of the plague developing in me. To this I agreed, and he promised to come for me next day. We talked on over the details of the provisioning and the defending of the Chemistry Building until the telephone died. It died in the midst of our conversation. That evening there were no electric lights, and I was alone in my house in the darkness. No more newspapers were being printed, so I had no knowledge of what was taking place outside.

"I heard sounds of rioting and of pistol shots, and from my windows I could see the glare of the sky of some conflagration in the direction of Oakland. It was a night of terror. I did not sleep a wink. A man—why and how I do not know—was killed on the sidewalk in front of the house. I heard the rapid reports of an automatic pistol, and a few minutes later the wounded wretch crawled up to my door, moaning and crying out for help. Arming myself with two automatics, I went to him. By the light of a match I ascertained that while he was dying of the bullet wounds, at the same time the plague was on him. I fled indoors, whence I heard him moan and cry out for half an hour longer.

"In the morning, my brother came to me. I had gathered into a handbag what things of value I purposed taking, but when I saw his face I knew that he would never accompany me to the Chemistry Building. The plague was on him. He intended shaking my hand, but I went back hurriedly before him.

"'Look at yourself in the mirror,' I commanded.

"He did so, and at sight of his scarlet face, the color deepening as he looked at it, he sank down nervelessly in a chair.

"'My God!' he said. 'I've got it. Don't come near me. I am a dead man.' Then the convulsions seized him. He was two hours in dying, and he was conscious to the last, complaining about the coldness and loss of sensation in his feet, his calves, his thighs, until at last it was his heart and he was dead.

"That was the way the Scarlet Death slew. I caught up my handbag and fled. The sights in the streets were terrible. One stumbled on bodies every-where. Some were not yet dead. And even as you looked, you saw men sink down with the death fastened upon them. There were numerous fires burning in Berkeley, while Oakland and San Francisco were apparently being swept by vast conflagrations. The smoke of the burning filled the heavens, so that the midday was as a gloomy twilight, and, in the shifts of wind, sometimes the sun shone through dimly, a dull red orb. Truly, my grandsons, it was like the last days of the end of the world.

"There were numerous stalled motor cars, showing that the gasoline and the engine supplies of the garages had given out. I remember one such car. A man and a woman lay back dead in the seats, and on the pavement near it were two more women and a child. Strange and terrible sights there were on every hand. People slipped by silently, furtively, like ghosts—white-faced women carrying infants in their arms; fathers leading children by the hand; singly, and in couples, and in families—all fleeing out of the city of death. Some carried supplies of food, others blankets and valuables, and there were many who carried nothing.

"There was a grocery store—a place where food was sold. The man to whom it belonged—I knew him well—a quiet, sober, but stupid and obstinate fel-low, was defending it. The windows and doors had been broken in, but he, inside, hiding behind a counter, was discharging his pistol at a number of men on the sidewalk who were breaking in. In the entrance were several bodies—of men, I decided, whom he had killed earlier in the day. Even as I looked on from a distance, I saw one of the robbers break the windows of the adjoining store, a place where shoes were sold, and deliberately set fire to it. I did not go to the groceryman's assistance. The time for such acts had already passed. Civilization was crumbling, and it was each for himself."

IV

"I WENT away hastily, down a cross-street, and at the first corner I saw another tragedy. Two men of the working class had caught a man and a woman with two children, and were robbing them. I knew the man by sight, though I had never been introduced to him. He was a poet whose verses I had long admired. Yet I did not go to his help, for at the moment I came upon the scene there was a pistol shot, and I saw him sinking to the ground. The woman screamed, and she was felled with a fist-blow by one of the brutes. I cried out threateningly, whereupon they discharged their pistols at me and I ran away around the corner. Here I was blocked by an advancing conflagration. The buildings on both sides were burning, and the street was filled with smoke and flame. From somewhere in that murk came a woman's voice calling shrilly for help. But I did not go to her. A man's heart turned to iron amid such scenes, and one heard all too many appeals for help.

"Returning to the corner, I found the two robbers were gone. The poet and his wife lay dead on the pavement. It was a shocking sight. The two children had vanished—whither I could not tell. And I knew, now, why it was that the fleeing persons I encountered slipped along so furtively and with such white faces. In the midst of our civilization, down in our slums and labor-ghettos, we had bred a race of barbarians, of savages; and now, in the time of our calamity, they turned upon us like the wild beasts they were and destroyed us. And they destroyed themselves as well.

"They inflamed themselves with strong drink and committed a thousand atrocities, quarreling and killing one another in the general madness. One group of workingmen I saw, of the better sort, who had banded together, and, with their women and children in their midst, the sick and aged in litters and being carried, and with a number of horses pulling a truck-load of provisions, they were fighting their way out of the city. They made a fine spectacle as they came down the street through the drifting smoke, though they nearly shot me when I first appeared in their path. As they went by, one of their leaders shouted out to me in apologetic explanation. He said they were killing the robbers and looters on sight, and that they had thus banded together as the only-means by which to escape the prowlers.

"It was here that I saw for the first time what I was soon to see so often. One of the marching men had suddenly shown the unmistakable mark of the plague. Immediately those about him drew away, and he, without a remonstrance, stepped out of his place to let them pass on. A woman, most probably his wife, attempted to follow him. She was leading a little boy by the hand. But the husband commanded her sternly to go on, while others laid hands on her and restrained her from following him. This I saw, and I saw the man also, with his scarlet blaze of face, step into a doorway on the opposite side of the street. I heard the report of his pistol, and saw him sink lifeless to the ground.

"After being turned aside twice again by advancing fires, I succeeded in getting through to the university. On the edge of the campus I came upon a party of university folk who were going in the direction of the Chemistry Building. They were all family men, and their families were with them, including the nurses and the servants. Professor Badminton greeted me, I had difficulty in recognizing him. Somewhere he had gone through flames, and his beard was singed off. About his head was a bloody bandage, and his clothes were filthy.

"He told me he had prowlers, and that his brother had been killed the previous night, in the defence of their dwelling.

"Midway across the campus, he pointed suddenly to Mrs. Swinton's face. The unmistakable scarlet was there. Immediately all the other women set up a screaming and began to run away from her. Her two children were with a nurse, and these also ran with the women. But her husband, Doctor Swinton, remained with her.

"'Go on, Smith,' he told me. 'Keep an eye on the children. As for me, I shall stay with my wife. I know she is as already dead, but I can't leave her. Afterwards, if I escape, I shall come to the Chemistry Building, and do you watch for me and let me in.'

"I left him bending over his wife and soothing her last moments, while I ran to overtake the party. We were the last to be admitted to the Chemistry Building. After that, with our automatic rifles we maintained our isolation. By our plans, we had arranged for a company of sixty to be in this refuge. Instead, every one of the number originally planned had added relatives and friends and whole families until there were over four hundred souls. But the Chemistry Building was large, and, standing by itself, was in no danger of being burned by the great fires that raged everywhere in the city.

"A large quantity of provisions had been gathered, and a food committee took charge of it, issuing rations daily to the various families and groups that arranged themselves into messes. A number of committees were appointed, and we developed a very efficient organization. I was on the committee of defence, though for the first day no prowlers came near. We could see them in the distance, however, and by the smoke of their fires knew that several camps of them were occupying the far edge of the campus. Drunkenness was rife, and often we heard them singing ribald songs or insanely shouting. While the world crashed to ruin about them and all the air was filled with the smoke of its burning, these low creatures gave rein to their bestiality and fought and drank and died. And after all, what did it matter? Everybody died anyway, the good and the bad, the efficients and the weaklings, those that loved to live and those that scorned to live. They passed. Everything passed. "When twenty-four hours had gone by and no signs of the plague were apparent, we congratulated ourselves and set about digging a well. You have seen the great iron pipes which in those days carried water to all the city-dwellers. We feared that the fires in the city would burst the pipes and empty the reservoirs. So we tore up the cement floor of the central court of the Chemistry Building and dug a well. There were many young men, undergraduates, with us, and we worked night and day on the well. And our fears were confirmed. Three hours before we reached water, the pipes went dry.

"A second twenty-four hours passed, and still the plague did not appear among us. We thought we were saved. But we did not know what I afterwards decided to be true, namely, that the period of the incubation of the plague germs in a human's body was a matter of a number of days. It slew so swiftly when once it manifested itself, that we were led to believe that the period of incubation was equally swift. So, when two days had left us unscathed, we were elated with the idea that we were free of the contagion.

"But the third day disillusioned us. I can never forget the night preceding it. I had charge of the night guards from eight to twelve, and from the roof of the building I watched the passing of all man's glorious works. So terrible were the local conflagrations that all the sky was lighted up. One could read the finest print in the red glare. All the world seemed wrapped in flames. San Francisco spouted smoke and fire from a score of vast conflagrations that were like so many active volcanoes. Oakland, San Leandro, Haywards—all were burning; and to the northward, clear to Point Richmond, other fires were at work. It was an awe-inspiring spectacle. Civilization, my grandsons, civilization was passing in a sheet of flame and a breath of death. At ten o'clock that night, the great powder magazines at Point Pinole exploded in rapid succession. So terrific were the concussions that the strong building rocked as in an earthquake, while every pane of glass was broken. It was then that I left the roof and went down the long corridors, from room to room, quieting the alarmed women and telling them what had happened.

"An hour later, at a window on the ground floor, I heard pandemonium break out in the camps of the prowlers. There were cries and screams, and shots from many pistols. As we afterward conjectured, this fight had been precipitated by an attempt on the part of those that were well to drive out those that were sick. At any rate, a number of the plague-stricken prowlers escaped across the campus and drifted against our doors. We warned them back, but they cursed us and discharged a fusillade from their pistols. Professor Merryweather, at one of the windows, was instantly killed, the bullet striking him squarely between the eyes. We opened fire in turn, and all the prowlers fled away with the exception of three. One was a woman. The plague was on them and they were reckless. Like foul fiends, there in the red glare from the skies, with faces blazing, they continued to curse us and fire at us. One of the men I shot with my own hand. After that the other man and the woman, still cursing us, lay down under our windows, where we were compelled to watch them die of the plague.

"The situation was critical. The explosions of the powder magazines had broken all the windows of the Chemistry Building, so that we were exposed to the germs from the corpses. The sanitary committee was called upon to act, and it responded nobly. Two men were required to go out and remove the corpses, and this meant the probable sacrifice of their own lives, for, having performed the task, they were not to be permitted to reenter the building. One of the professors, who was a bachelor, and one of the undergraduates volunteered. They bade good-bye to us and went forth. They were heroes. They gave up their lives that four hundred others might live. After they had performed their work, they stood for a moment, at a distance, looking at us wistfully. Then they waved their hands in farewell and went away slowly across the campus toward the burning city.

"And yet it was all useless. The next morning the first one of us was smitten with the plague—a little nurse-girl in the family of Professor Stout. It was no time for weak-kneed, sentimental policies. On the chance that she might be the only one, we thrust her forth from the building and commanded her to be gone.

"She went away slowly across the campus, wringing her hands and crying pitifully. We felt like brutes, but what were we to do? There were four hundred of us, and individuals had to be sacrificed.

"In one of the laboratories three families had domiciled themselves, and that afternoon we found among them no less than four corpses and seven cases of the plague in all its different stages.

"Then it was that the horror began. Leaving the dead lie, we forced the living ones to segregate themselves in another room. The plague began to break out among the rest of us, and as fast as the symptoms appeared, we sent the stricken ones to these segregated rooms. We compelled them to walk there by themselves, so as to avoid laying hands on them. It was heartrending. But still the plague raged among us, and room after room was filled with the dead and dying. And so we who were yet clean retreated to the next floor and to the next, before this sea of the dead, that, room by room and floor by floor, inundated the building.

"The place became a charnel house, and in the middle of the night the survivors fled forth, taking nothing with them except arms and ammunition and a heavy store of tinned foods. We camped on the opposite side of the campus from the prowlers, and, while some stood guard, others of us volunteered to scout into the city in quest of horses, motor cars, carts, and wagons, or anything that would carry our provisions and enable us to emulate the banded workingmen I had seen fighting their way out to the open country.

"I was one of these scouts; and Doctor Hoyle, remembering that his motor car had been left behind in his home garage, told me to look for it. We scouted in pairs, and Dombey, a young undergraduate, accompanied me. We had to cross half a mile of the residence portion of the city to get to Doctor Hoyle's home. Here the buildings stood apart, in the midst of trees and grassy lawns, and here the fires had played freaks, burning whole blocks, skipping blocks and often skipping a single house in a block. And here, too, the prowlers were still at their work. We carried our automatic pistols openly in our hands, and looked desperate enough, forsooth, to keep them from attacking us. But at Doctor Hoyle's house the thing happened. Untouched by fire, even as we came to it the smoke of flames burst forth.

"The miscreant who had set fire to it staggered down the steps and out along the driveway. Sticking out of his coat pockets were bottles of whiskey, and he was very drunk. My first impulse was to shoot him, and I have never ceased regretting that I did not. Staggering and maundering to himself, with bloodshot eyes, and a raw and bleeding slash down one side of his bewhiskered face, he was altogether the most nauseating specimen of degradation and filth I had ever encountered. I did not shoot him, and he leaned against a tree on the lawn to let us go by. It was the most absolute, wanton act. Just as we were opposite him, he suddenly drew a pistol and shot Dombey through the head. The next instant I shot him. But it was too late. Dombey expired without a groan, immediately. I doubt if he even knew what had happened to him.

"Leaving the two corpses, I hurried on past the burning house to the garage, and there found Doctor Hoyle's motor car. The tanks were filled with gasoline, and it was ready for use. And it was in this car that I threaded the streets of the ruined city and came back to the survivors on the campus. The other scouts returned, but none had been so fortunate. Professor Fairmead had found a Shetland pony, but the poor creature, tied in a stable and abandoned for days, was so weak from want of food and water that it could carry no burden at all. Some of the men were for turning it loose, but I insisted that we should lead it along with us, so that, if we got out of food, we would have it to eat.

"There were forty-seven of us when we started, many being women and children. The President of the Faculty, an old man to begin with, and now hopelessly broken by the awful happenings of the past week, rode in the motor car with several young children and the aged mother of Professor Fairmead. Wathope, a young professor of English, who had a grievous bullet-wound in his leg, drove the car. The rest of us walked, Professor Fairmead leading the pony.

"It was what should have been a bright summer day, but the smoke from the burning world filled the sky, through which the sun shone murkily, a dull and lifeless orb, blood-red and ominous. But we had grown accustomed to that blood-red sun. With the smoke it was different. It bit into our nostrils and eyes, and there was not one of us whose eyes were not bloodshot. We directed our course to the southeast through the endless miles of suburban residences, travelling along where the first swells of low hills rose from the flat of the central city. It was by this way, only, that we could expect to gain the country.

"Our progress was painfully slow. The women and children could not walk fast. They did not dream of walking, my grandsons, in the way all people walk today. In truth, none of us knew how to walk. It was not until after the plague that I learned really to walk. So it was that the pace of the slowest was the pace of all, for we dared not separate on account of the prowlers. There were not so many now of these human beasts of prey. The plague had already well diminished their numbers, but enough still lived to be a constant menace to us. Many of the beautiful residences were untouched by fire, yet smoking ruins were everywhere. The prowlers, too, seemed to have got over their insensate desire to burn, and it was more rarely that we saw houses freshly on fire.

"Several of us scouted among the private garages in search of motor cars and gasoline. But in this we were unsuccessful. The first great flights from the cities had swept all such utilities away. Calgan, a fine young man, was lost in this work. He was shot by prowlers while crossing a lawn. Yet this was our only casualty, though, once, a drunken brute deliberately opened fire on all of us. Luckily, he fired wildly, and we shot him before he had done any hurt.

"At Fruitvale, still in the heart of the magnificent residence section of the city, the plague again smote us. Professor Fairmead was the victim. Making signs to us that his mother was not to know, he turned aside into the grounds of a beautiful mansion. He sat down forlornly on the steps of the front veranda, and I, having lingered, waved him a last farewell. That night, several miles beyond Fruitvale and still in the city, we made camp. And that night we shifted camp twice to get away from our dead. In the morning there were thirty of us. I shall never forget the President of the Faculty. During the morning's march his wife, who was walking, betrayed the fatal symptoms, and when she drew aside to let us go on, he insisted on leaving the motor car and remaining with her. There was quite a discussion about this, but in the end we gave in. It was just as well, for we knew not which ones of us, if any, might ultimately escape.

"That night, the second of our march, we camped beyond Haywards in the first stretches of country. And in the morning there were eleven of us that lived. Also, during the night, Wathope, the professor with the wounded leg, deserted us in the motor car. He took with him his sister and his mother and most of our tinned provisions. It was that day, in the afternoon, while resting by the wayside, that I saw the last airship I shall ever see. The smoke was much thinner here in the country, and I first sighted the ship drifting and veering helplessly at an elevation of two thousand feet. What had happened I could not conjecture, but even as we looked we saw her bow dip down lower and lower. Then the bulkheads of the various gas-chambers must have burst, for, quite perpendicular, she fell like a plummet to the earth.

"And from that day to this I have not seen another airship. Often and often, during the next few years, I scanned the sky for them, hoping against hope that somewhere in the world civilization had survived. But it was not to be. What happened with us in California must have happened with everybody everywhere.

"Another day, and at Niles there were three of us. Beyond Niles, in the middle of the highway, we found Wathope. The motor car had broken down, and there, on the rugs which they had spread on the ground, lay the bodies of his sister, his mother, and himself.

"Wearied by the unusual exercise of continual walking, that night I slept heavily. In the morning I was alone in the world. Canfield and Parsons, my last companions, were dead of the plague. Of the four hundred that sought shelter in the Chemistry Building, and of the forty-seven that began the march, I alone remained—I and the Shetland pony. Why this should be so there is no explaining. I did not catch the plague, that is all. I was immune. I was merely the one lucky man in a million—just as every survivor was one in a million, or, rather, in several millions, for the proportion was at least that."

V

"FOR two days I sheltered in a pleasant grove where there had been no deaths. In those two days, while badly depressed and believing that my turn would come at any moment, nevertheless I rested and recuperated. So did the pony.

And on the third day, putting what small store of tinned provisions I possessed on the pony's back, I started on across a very lonely land. Not a live man, woman, or child, did I encounter, though the dead were everywhere. Food, however, was abundant. The land then was not as it is now. It was all cleared of trees and brush, and it was cultivated. The food for millions of mouths was growing, ripening, and going to waste. From the fields and orchards I gathered vegetables, fruits, and berries. Around the deserted farmhouses I got eggs and caught chickens. And frequently I found supplies of tinned provisions in the store-rooms.

"A strange thing was what was taking place with all the domestic animals. Everywhere they were going wild and preying on one another. The chickens and ducks were the first to be destroyed, while the pigs were the first to go wild, followed by the cats. Nor were the dogs long in adapting themselves to the changed conditions. There was a veritable plague of dogs. They devoured the corpses, barked and howled during the nights, and in the daytime slunk about in the distance. As the time went by, I noticed a change in their behavior. At first they were apart from one another, very suspicious and very prone to fight. But after a not very long while they began to come together and run in packs. The dog, you see, always was a social animal, and this was true before ever he came to be domesticated by man. In the last days of the world before the plague, there were many many very different kinds of dogs—dogs without hair and dogs with warm fur, dogs so small that they would make scarcely a mouthful for other dogs that were as large as mountain lions. Well, all the small dogs, and the weak types, were killed by their fellows. Also, the very large ones were not adapted for the wild life and bred out. As a result, the many different kinds of dogs disappeared, and there remained, running in packs, the medium-sized wolfish dogs that you know today."

"But the cats don't run in packs, Granser," Hoo-Hoo objected.

"The cat was never a social animal. As one writer in the nineteenth century said, the cat walks by himself. He always walked by himself, from before the time he was tamed by man, down through the long ages of domestication, to today when once more he is wild.

"The horses also went wild, and all the fine breeds we had degenerated into the small mustang horse you know today. The cows likewise went wild, as did the pigeons and the sheep. And that a few of the chickens survived you know yourself. But the wild chicken of today is quite a different thing from the chickens we had in those days.

"But I must go on with my story. I travelled through a deserted land. As the time went by I began to yearn more and more for human beings. But I never found one, and I grew lonelier and lonelier. I crossed Livermore Valley and the mountains between it and the great valley of the San Joaquin. You have never seen that valley, but it is very large and it is the home of the wild horse. There are great droves there, thousands and tens of thousands. I revisited it thirty years after, so I know. You think there are lots of wild horses down here in the coast valleys, but they are as nothing compared with those of the San Joaquin. Strange to say, the cows, when they went wild, went back into the lower mountains. Evidently they were better able to protect themselves there. "In the country districts the ghouls and prowlers had been less in evidence, for I found many villages and towns untouched by fire. But they were filled by the pestilential dead, and I passed by without exploring them. It was near Lathrop that, out of my loneliness, I picked up a pair of collie dogs that were so newly free that they were urgently willing to return to their allegiance to man. These collies accompanied me for many years, and the strains of them are in those very dogs there that you boys have today. But in sixty years the collie strain has worked out. These brutes are more like domesticated wolves than anything else."

Hare-Lip rose to his feet, glanced to see that the goats were safe, and looked at the sun's position in the afternoon sky, advertising impatience at the prolixity of the old man's tale. Urged to hurry by Edwin, Granser went on. "There is little more to tell. With my two dogs and my pony, and riding a horse I had managed to capture, I crossed the San Joaquin and went on to a wonderful valley in the Sierras called Yosemite. In the great hotel there I found a prodigious supply of tinned provisions. The pasture was abundant, as was the game, and the river that ran through the valley was full of trout. I remained there three years in an utter loneliness that none but a man who has once been highly civilized can understand. Then I could stand it no more. I felt that I was going crazy. Like the dog, I was a social animal and I needed my kind. I reasoned that since I had survived the plague, there was a possibility that others had survived. Also, I reasoned that after three years the plague germs must all be gone and the land be clean again.

"With my horse and dogs and pony, I set out. Again I crossed the San Joaquin Valley, the mountains beyond, and came down into Livermore Valley. The change in those three years was amazing. All the land had been splendidly tilled, and now I could scarcely recognize it, such was the sea of rank vegetation that had overrun the agricultural handiwork of man. You see, the wheat, the vegetables, and orchard trees had always been cared for and nursed by man, so that they were soft and tender. The weeds and wild bushes and such things, on the contrary, had always been fought by man, so that they were tough and resistant. As a result, when the hand of man was removed, the wild vegetation smothered and destroyed practically all the domesticated vegetation. The coyotes were greatly increased, and it was at this time that I first encountered wolves, straying in twos and threes and small packs down from the regions where they had always persisted.

"It was at Lake Temescal, not far from the one-time city of Oakland, that I came upon the first live human beings. Oh, my grandsons, how can I describe to you my emotion, when, astride my horse and dropping down the hillside to the lake, I saw the smoke of a campfire rising through the trees. Almost did my heart stop beating. I felt that I was going crazy. Then I heard the cry of a babe—a human babe. And dogs barked, and my dogs answered. I did not know but what I was the one human alive in the whole world. It could not be true that here were others—smoke, and the cry of a babe.

"Emerging on the lake, there, before my eyes, not a hundred yards away, I saw a man, a large man. He was standing on an outjutting rock and fishing. I was overcome. I stopped my horse. I tried to call out but could not. I waved my hand. It seemed to me that the man looked at me, but he did not appear to wave. Then I laid my head on my arms there in the saddle. I was afraid to look again, for I knew it was an hallucination, and I knew that if I looked the man would be gone. And so precious was the hallucination, that I wanted it to persist yet a little while. I knew, too, that as long as I did not look it would persist.

"Thus I remained, until I heard my dogs snarling, and a man's voice. What do you think the voice said? I will tell you. It said: '*Where in hell did you come from?*'

"Those were the words, the exact words. That was what your other grandfather said to me, Hare-Lip, when he greeted me there on the shore of Lake Temescal fifty-seven years ago. And they were the most ineffable words I have ever heard. I opened my eyes, and there he stood before me, a large, dark, hairy man, heavy-jawed, slant-browed, fierce-eyed. How I got off my horse I do not know. But it seemed that the next I knew I was clasping his hand with both of mine and crying. I would have embraced him, but he was ever a narrow-minded, suspicious man, and he drew away from me. Yet did I cling to his hand and cry."

Granser's voice faltered and broke at the recollection, and the weak tears streamed down his cheeks while the boys looked on and giggled.

"Yet did I cry," he continued, "and desire to embrace him, though the Chauffeur was a brute, a perfect brute—the most abhorrent man I have ever known. His name was… strange, how I have forgotten his name. Everybody called him Chauffeur—it was the name of his occupation, and it stuck. That is how, to this day, the tribe he founded is called the Chauffeur Tribe.

"He was a violent, unjust man. Why the plague germs spared him I can never understand. It would seem, in spite of our old metaphysical notions about absolute justice, that there is no justice in the universe. Why did he live?—an iniquitous, moral monster, a blot on the face of nature, a cruel, relentless, bestial cheat as well. All he could talk about was motor cars, machinery, gasoline, and garages—and especially, and with huge delight, of his mean pilferings and sordid swindlings of the persons who had employed him in the days before the coming of the plague. And yet he was spared, while hundreds of millions, yea, billions, of better men were destroyed.

"I went on with him to his camp, and there I saw her, Vesta, the one woman. It was glorious and… pitiful. There she was, Vesta Van Warden, the young wife of John Van Warden, clad in rags, with marred and scarred and toil-calloused hands, bending over the campfire and doing scullion work—she, Vesta, who had been born to the purple of the greatest baronage of wealth the world had ever known. John Van Warden, her husband, worth one billion, eight hundred millions and President of the Board of Industrial Magnates, had been the ruler of America. Also, sitting on the International Board of Control, he had been one of the seven men who ruled the world. And she herself had come of equally noble stock. Her father, Philip Saxon, had been President of the Board of Industrial Magnates up to the time of his death. This office was in process of becoming hereditary, and had Philip Saxon had a son that son would have succeeded him. But his only child was Vesta, the perfect flower of generations of the highest culture this planet has ever produced. It was not until the engagement between Vesta and Van Warden took place, that Saxon indicated the latter as his successor. It was, I am sure, a political marriage. I have reason to believe that Vesta never really loved her husband in the mad passionate way of which the poets used to sing. It was more like the marriages that obtained among crowned heads in the days before they were displaced by the Magnates.

"And there she was, boiling fish-chowder in a soot-covered pot, her glorious eyes inflamed by the acrid smoke of the open fire. Hers was a sad story. She was the one survivor in a million, as I had been, as the Chauffeur had been. On a crowning eminence of the Alameda Hills, overlooking San Francisco Bay, Van Warden had built a vast summer palace. It was surrounded by a park of a thousand acres. When the plague broke out, Van Warden sent her there. Armed guards patrolled the boundaries of the park, and nothing entered in the way of provisions or even mail matter that was not first fumigated. And yet did the plague enter, killing the guards at their posts, the servants at their tasks, sweeping away the whole army of retainers—or, at least, all of them who did not flee to die elsewhere. So it was that Vesta found herself the sole living person in the palace that had become a charnel house.

"Now the Chauffeur had been one of the servants that ran away. Returning, two months afterward, he discovered Vesta in a little summer pavilion where there had been no deaths and where she had established herself. He was a brute. She was afraid, and she ran away and hid among the trees. That night, on foot, she fled into the mountains—she, whose tender feet and delicate body had never known the bruise of stones nor the scratch of briars. He followed, and that night he caught her. He struck her. Do you understand? He beat her with those terrible fists of his and made her his slave. It was she who had to gather the firewood, build the fires, cook, and do all the degrading camp-labor—she, who had never performed a menial act in her life. These things he compelled her to do, while he, a proper savage, elected to lie around camp and look on. He did nothing, absolutely nothing, except on occasion to hunt meat or catch fish."

"Good for Chauffeur," Hare-Lip commented in an undertone to the other boys. "I remember him before he died. He was a corker. But he did things, and he made things go. You know, Dad married his daughter, an' you ought to see the way he knocked the spots outa Dad. The Chauffeur was a son-of-a-gun. He made us kids stand around. Even when he was croaking he reached out for me, once, an' laid my head open with that long stick he kept always beside him."

Hare-Lip rubbed his bullet head reminiscently, and the boys returned to the old man, who was maundering ecstatically about Vesta, the squaw of the founder of the Chauffeur Tribe.

"And so I say to you that you cannot understand the awfulness of the situation. The Chauffeur was a servant, understand, a servant. And he cringed, with bowed head, to such as she. She was a lord of life, both by birth and by marriage. The destinies of millions, such as he, she carried in the hollow of her pink-white hand. And, in the days before the plague, the slightest contact with such as he would have been pollution. Oh, I have seen it. Once, I remember, there was Mrs. Goldwin, wife of one of the great magnates. It was on a landing stage, just as she was embarking in her private dirigible, that she dropped her parasol. A servant picked it up and made the mistake of handing it to her—to her, one of the greatest royal ladies of the land! She shrank back, as though he were a leper, and indicated her secretary to receive it. Also, she ordered her secretary to ascertain the creature's name and to see that he was immediately discharged from service. And such a woman was Vesta Van Warden. And her the Chauffeur beat and made his slave.

"—Bill—that was it; Bill, the Chauffeur. That was his name. He was a wretched, primitive man, wholly devoid of the finer instincts and chivalrous promptings of a cultured soul. No, there is no absolute justice, for to him fell that wonder of womanhood, Vesta Van Warden. The grievousness of this you will never understand, my grandsons; for you are yourselves primitive little savages, unaware of aught else but savagery. Why should Vesta not have been mine? I was a man of culture and refinement, a professor in a great university. Even so, in the time before the plague, such was her exalted position, she would not have deigned to know that I existed. Mark, then, the abysmal degradation to which she fell at the hands of the Chauffeur. Nothing less than the destruction of all mankind had made it possible that I should know her, look in her eyes, converse with her, touch her hand—ay, and love her and know that her feelings toward me were very kindly. I have reason to believe that she, even she, would have loved me, there being no other man in the world except the Chauffeur. Why, when it destroyed eight billions of souls, did not the plague destroy just one more man, and that man the Chauffeur?

"Once, when the Chauffeur was away fishing, she begged me to kill him. With tears in her eyes she begged me to kill him. But he was a strong and violent man, and I was afraid. Afterwards, I talked with him. I offered him my horse, my pony, my dogs, all that I possessed, if he would give Vesta to me. And he grinned in my face and shook his head. He was very insulting. He said that in the old days he had been a servant, had been dirt under the feet of men like me and of women like Vesta, and that now he had the greatest lady in the land to be servant to him and cook his food and nurse his brats. 'You had your day before the plague,' he said; 'but this is my day, and a damned good day it is. I wouldn't trade back to the old times for anything.' Such words he spoke, but they are not his words. He was a vulgar, low-minded man, and vile oaths fell continually from his lips.

"Also, he told me that if he caught me making eyes at his woman he'd wring my neck and give her a beating as well. What was I to do? I was afraid. He was a brute. That first night, when I discovered the camp, Vesta and I had great talk about the things of our vanished world. We talked of art, and books, and poetry; and the Chauffeur listened and grinned and sneered. He was bored and angered by our way of speech which he did not comprehend, and finally he spoke up and said: 'And this is Vesta Van Warden, one-time wife of Van Warden the Magnate—a high and stuck-up beauty, who is now my squaw. Eh, Professor Smith, times is changed, times is changed. Here, you, woman, take off my moccasins, and lively about it. I want Professor Smith to see how well I have you trained.'

"I saw her clench her teeth, and the flame of revolt rise in her face. He drew back his gnarled fist to strike, and I was afraid, and sick at heart. I could do nothing to prevail against him. So I got up to go, and not be witness to such indignity. But the Chauffeur laughed and threatened me with a beating if I did not stay and behold. And I sat there, perforce, by the campfire on the shore of Lake Temescal, and saw Vesta, Vesta Van Warden, kneel and remove the moccasins of that grinning, hairy, apelike human brute.

"—Oh, you do not understand, my grandsons. You have never known anything else, and you do not understand.

"'Halter-broke and bridle-wise,' the Chauffeur gloated, while she per-formed that dreadful, menial task. 'A trifle balky at times, Professor, a trifle balky; but a clout alongside the jaw makes her as meek and gentle as a lamb.'

"And another time he said: 'We've got to start all over and replenish the earth and multiply. You're handicapped, Professor. You ain't got no wife, and we're up against a regular Garden-of-Eden proposition. But I ain't proud. I'll tell you what, Professor.' He pointed at their little infant, barely a year old. 'There's your wife, though you'll have to wait till she grows up. It's rich, ain't it? We're all equals here, and I'm the biggest toad in the splash. But I ain't stuck up—not I. I do you the honor, Professor Smith, the very great honor of betrothing to you my and Vesta Van Warden's daughter. Ain't it cussed bad that Van Warden ain't here to see?'"

VI

"I LIVED three weeks of infinite torment there in the Chauffeur's camp. And then, one day, tiring of me, or of what to him was my bad effect on Vesta, he told me that the year before, wandering through the Contra Costa Hills to the Straits of Carquinez, across the Straits he had seen a smoke. This meant that there were still other human beings, and that for three weeks he had kept this inestimably precious information from me. I departed at once, with my dogs and horses, and journeyed across the Contra Costa Hills to the Straits. I saw no smoke on the other side, but at Port Costa discovered a small steel barge on which I was able to embark my animals. Old canvas which I found served me for a sail, and a southerly breeze fanned me across the Straits and up to the ruins of Vallejo. Here, on the outskirts of the city, I found evidences of a recently occupied camp.

"Many clam-shells showed me why these humans had come to the shores of the Bay. This was the Santa Rosa Tribe, and I followed its track along the old railroad right of way across the salt marshes to Sonoma Valley. Here, at the old brickyard at Glen Ellen, I came upon the camp. There were eighteen souls all told. Two were old men, one of whom was Jones, a banker. The other was

Harrison, a retired pawnbroker, who had taken for wife the matron of the State Hospital for the Insane at Napa. Of all the persons of the city of Napa, and of all the other towns and villages in that rich and populous valley, she had been the only survivor. Next, there were the three young men—Cardiff and Hale, who had been farmers, and Wainwright, a common day-laborer. All three had found wives. To Hale, a crude, illiterate farmer, had fallen Isadore, the greatest prize, next to Vesta, of the women who came through the plague. She was one of the world's most noted singers, and the plague had caught her at San Francisco. She has talked with me for hours at a time, telling me of her adventures, until, at last, rescued by Hale in the Mendocino Forest Reserve, there had remained nothing for her to do but become his wife. But Hale was a good fellow, in spite of his illiteracy. He had a keen sense of justice and right-dealing, and she was far happier with him than was Vesta with the Chauffeur.

"The wives of Cardiff and Wainwright were ordinary women, accustomed to toil with strong constitutions—just the type for the wild new life which they were compelled to live. In addition were two adult idiots from the feeble-minded home at Eldredge, and five or six young children and infants born after the formation of the Santa Rosa Tribe. Also, there was Bertha. She was a good woman, Hare-Lip, in spite of the sneers of your father. Her I took for wife. She was the mother of your father, Edwin, and of yours, Hoo-Hoo. And it was our daughter, Vera, who married your father, Hare-Lip—your father, Sandow, who was the oldest son of Vesta Van Warden and the Chauffeur.

"And so it was that I became the nineteenth member of the Santa Rosa Tribe. There were only two outsiders added after me. One was Mungerson, descended from the Magnates, who wandered alone in the wilds of Northern California for eight years before he came south and joined us. He it was who waited twelve years more before he married my daughter, Mary. The other was Johnson, the man who founded the Utah Tribe. That was where he came from, Utah, a country that lies very far away from here, across the great deserts, to the east. It was not until twenty-seven years after the plague that Johnson reached California. In all that Utah region he reported but three survivors, himself one, and all men. For many years these three men lived and hunted together, until, at last, desperate, fearing that with them the human race would perish utterly from the planet, they headed westward on the possibility of finding women survivors in California. Johnson alone came through the great desert, where his two companions died. He was forty-six years old when he joined us, and he married the fourth daughter of Isadore and Hale, and his eldest son married your aunt, Hare-Lip, who was the third daughter of Vesta and the Chauffeur. Johnson was a strong man, with a will of his own. And it was because of this that he seceded from the Santa Rosans and formed the Utah Tribe at San José. It is a small tribe—there are only nine in it; but, though he is dead, such was his influence and the strength of his breed, that it will grow into a strong tribe and play a leading part in the recivilization of the planet.

"There are only two other tribes that we know of—the Los Angelitos and the Carmelitos. The latter started from one man and woman. He was called Lopez, and he was descended from the ancient Mexicans and was very black. He was a cowherd in the ranges beyond Carmel, and his wife was a maidservant in the great Del Monte Hotel. It was seven years before we first got in touch with the Los Angelitos. They have a good country down there, but it is too warm. I estimate the present population of the world at between three hundred and fifty and four hundred—provided, of course, that there are no scattered little tribes elsewhere in the world. If there be such, we have not heard from them. Since Johnson crossed the desert from Utah, no word nor sign has come from the East or anywhere else. The great world which I knew in my boyhood and early manhood is gone. It has ceased to be. I am the last man who was alive in the days of the plague and who knows the wonders of that far-off time. We, who mastered the planet—its earth, and sea, and sky—and who were as very gods, now live in primitive savagery along the water courses of this California country.

"But we are increasing rapidly—your sister, Hare-Lip, already has four children. We are increasing rapidly and making ready for a new climb toward civilization. In time, pressure of population will compel us to spread out, and a hundred generations from now we may expect our descendants to start across the Sierras, oozing slowly along, generation by generation, over the great continent to the colonization of the East—a new Aryan drift around the world.

"But it will be slow, very slow; we have so far to climb. We fell so hopelessly far. If only one physicist or one chemist had survived! But it was not to be, and we have forgotten everything. The Chauffeur started working in iron. He made the forge which we use to this day. But he was a lazy man, and when he died he took with him all he knew of metals and machinery. What was I to know of such things? I was a classical scholar, not a chemist. The other men who survived were not educated. Only two things did the Chauffeur accomplish—the brewing of strong drink and the growing of tobacco. It was while he was drunk, once, that he killed Vesta. I firmly believe that he killed Vesta in a fit of drunken cruelty though he always maintained that she fell into the lake and was drowned.

"And, my grandsons, let me warn you against the medicine-men. They call themselves *doctors*, travestying what was once a noble profession, but in reality they are medicine-men, devil-devil men, and they make for superstition and darkness. They are cheats and liars. But so debased and degraded are we, that we believe their lies. They, too, will increase in numbers as we increase, and they will strive to rule us. Yet are they liars and charlatans. Look at young Cross-Eyes, posing as a doctor, selling charms against sickness, giving good hunting, exchanging promises of fair weather for good meat and skins, sending the death-stick, performing a thousand abominations. Yet I say to you, that when he says he can do these things, he lies. I, Professor Smith, Professor James Howard Smith, say that he lies. I have told him so to his teeth. Why has he not sent me the death-stick? Because he knows that with me it is without avail. But you, Hare-Lip, so deeply are you sunk in black superstition that did you awake this night and find the death-stick beside you, you would surely die. And you would die, not because of any virtues in the stick, but because you are a savage with the dark and clouded mind of a savage.

"The doctors must be destroyed, and all that was lost must be discovered over again. Wherefore, earnestly, I repeat unto you certain things which you must remember and tell to your children after you. You must tell them that when water is made hot by fire, there resides in it a wonderful thing called steam, which is stronger than ten thousand men and which can do all man's work for him. There are other very useful things. In the lightning flash resides a similarly strong servant of man, which was of old his slave and which some day will be his slave again.

"Quite a different thing is the alphabet. It is what enables me to know the meaning of fine markings, whereas you boys know only rude picture-writing. In that dry cave on Telegraph Hill, where you see me often go when the tribe is down by the sea, I have stored many books. In them is great wisdom. Also, with them, I have placed a key to the alphabet, so that one who knows picture-writing may also know print. Some day men will read again; and then, if no accident has befallen my cave, they will know that Professor James Howard Smith once lived and saved for them the knowledge of the ancients. "There is another little device that men inevitably will rediscover. It is called gunpowder. It was what enabled us to kill surely and at long distances. Certain things which are found in the ground, when combined in the right proportions, will make this gunpowder. What these things are, I have forgotten, or else I never knew. But I wish I did know. Then would I make powder, and then would I certainly kill Cross-Eyes and rid the land of superstition—"

"After I am man-grown I am going to give Cross-Eyes all the goats, and meat, and skins I can get, so that he'll teach me to be a doctor," Hoo-Hoo asserted. "And when I know, I'll make everybody else sit up and take notice. They'll get down in the dirt to me, you bet."

The old man nodded his head solemnly, and murmured:

"Strange it is to hear the vestiges and remnants of the complicated Aryan speech falling from the lips of a filthy little skin-clad savage. All the world is topsy-turvy. And it has been topsy-turvy ever since the plague."

"You won't make me sit up," Hare-Lip boasted to the would-be medicine-man. "If I paid you for a sending of the death-stick and it didn't work, I'd bust in your head—understand, you Hoo-Hoo, you?"

"I'm going to get Granser to remember this here gunpowder stuff," Edwin said softly, "and then I'll have you all on the run. You, Hare-Lip, will do my fighting for me and get my meat for me, and you, Hoo-Hoo, will send the death-stick for me and make everybody afraid. And if I catch Hare-Lip trying to bust your head, Hoo-Hoo, I'll fix him with that same gunpowder. Granser ain't such a fool as you think, and I'm going to listen to him and some day I'll be boss over the whole bunch of you."

The old man shook his head sadly, and said:

"The gunpowder will come. Nothing can stop it—the same old story over and over. Man will increase, and men will fight. The gunpowder will enable men to kill millions of men, and in this way only, by fire and blood, will a new civilization, in some remote day, be evolved. And of what profit will it be? Just as the old civilization passed, so will the new. It may take fifty thousand years to build, but it will pass. All things pass. Only remain cosmic force and matter, ever in flux, ever acting and reacting and realizing the eternal types—the priest, the soldier, and the king. Out of the mouths of babes comes the wisdom of all the ages. Some will fight, some will rule, some will pray; and all the rest will toil and suffer sore while on their bleeding carcasses is reared again, and yet again, without end, the amazing beauty and surpassing wonder of the civilized state. It were just as well that I destroyed those cave-stored books—whether they remain or perish, all their old truths will be discovered, their old lies lived and handed down. What is the profit—"

Hare-Lip leaped to his feet, giving a quick glance at the pasturing goats and the afternoon sun.

"Gee!" he muttered to Edwin, "The old geezer gets more long-winded every day. Let's pull for camp."

While the other two, aided by the dogs, assembled the goats and started them for the trail through the forest, Edwin stayed by the old man and guided him in the same direction. When they reached the old right of way, Edwin stopped suddenly and looked back. Hare-Lip and Hoo-Hoo and the dogs and the goats passed on. Edwin was looking at a small herd of wild horses which had come down on the hard sand. There were at least twenty of them, young colts and yearlings and mares, led by a beautiful stallion which stood in the foam at the edge of the surf, with arched neck and bright wild eyes, sniffing the salt air from off the sea.

"What is it?" Granser queried.

"Horses," was the answer. "First time I ever seen 'em on the beach. It's the mountain lions getting thicker and thicker and driving 'em down."

The low sun shot red shafts of light, fan-shaped, up from a cloud-tumbled horizon. And close at hand, in the white waste of shore-lashed waters, the sea-lions, bellowing their old primeval chant, hauled up out of the sea on the black rocks and fought and loved.

"Come on, Granser," Edwin prompted. And old man and boy, skin-clad and barbaric, turned and went along the right of way into the forest in the wake of the goats.

THE END

John Griffith "Jack" London (1876–1916) was an American novelist, one of the first fiction writers to get rich from writing fiction. His most famous novels— *The Call of the Wild* (1903) and *White Fang* (1906)—were adventure stories set in the Klondike Gold Rush. A number of London's stories were science fiction, and dealt with topics such as invisibility, germ warfare, and energy weapons.

Commentary

In 1918, just six years after "The Scarlet Plague" was published in *London Magazine*, the deadly Spanish flu pandemic struck humanity. (The disease got its name not because it originated in Spain but simply because reporters were free to describe its dread effects there. In order to maintain morale, wartime censors suppressed early reports of the illness in the nations fighting World War I. At the time, therefore, it seemed as if neutral Spain had been particularly badly affected.) The H1N1 influenza virus circulating in the period January 1918 to December 1920 infected half a billion people around the globe and killed as many as 100 million people—five percent of the human population. Flu deaths far exceeded the number killed in four years of battle. The disease struck everywhere, from the Arctic to isolated islands in the Pacific. In some communities the impact was so terrible survivors decided it would be best to never talk about it—to pretend, as it were, that this grim event had simply not happened.

Spanish flu was merely the most recent of large-scale pandemics; infectious disease has attacked humankind throughout history. Bubonic plague has been responsible for the most notable pandemics. The first recorded outbreak was the Plague of Justinian (541–542), which struck the Byzantine Empire. Historians estimate that 25 million people died of the plague—less than the number of fatalities caused by Spanish flu, but the total population was much smaller back in Roman times. The Plague of Justinian killed about 13% of the world's population. The plague returned over the following two centuries, killing a further 25 million people.

The second major outbreak of plague began with the Black Death in 1347. The disease made numerous returns over the following three centuries. The Black Death was one of the most devastating events in human history: it killed between 75 and 200 million people. The dead were placed in ditches and then isolated; in some towns and villages there weren't enough living to bury the dead. (Figure 3.1 shows an example of a plague pit.) The world population did not recover to pre-Black Death levels until the seventeenth century. Many historians argue that the devastation unleashed by this pandemic had a significant impact on the course of European history: the labour shortages it caused accelerated several economic, social, and technical developments, and might even have helped introduce the Renaissance.

So we know pandemics occur. It's entirely possible—even probable—that a destructive flu pandemic will strike again. And it's not hard to imagine how a mutation in one of the viral hemorrhagic fevers—Ebola, say, or the Marburg virus—could lead to a disease causing widespread suffering. Therefore the basic premise of "The Scarlet Plague" is not unrealistic. But Jack London was writing more than a century ago. Although elements of his story were prescient (for example, he guessed a global population of eight billion people by 2010—not bad), science and technology have advanced to a level he could scarcely have imagined. Suppose a disease such as the scarlet plague did break out, and let's assume it was as lethal as, say, Spanish flu. Would our modern civilization collapse in the way London suggests?

This is not an easy question to answer.

On the one hand, as I write, the world population is about 7.6 billion. A large fraction of the population is mobile to an extent that would have astonished Jack London. In the past, disease travelled from continent to continent at the speed of sailing boats. Nowadays, if people develop a disease in Beijing, say, they can carry it to Berlin within hours. This combination of a large pool of people in which disease can develop with the rapid, large-scale movement of people mean infections can be transmitted more efficiently than ever before. Furthermore, the threats posed by viruses and bacteria are always evolving. The flu virus, for example, changes constantly. One type of change is a gradual "drift" in genetic make-up that leads, over time, to a virus our immune system

Fig. 3.1 These are believed to be the skulls and bones of people who succumbed to the plague, although usually in such cases skeletons are found in some semblance of order. The bodies of these unfortunates were dumped in a brick-built pit (Credit: Wellcome Collection gallery)

fails to recognise even if we have previously been exposed to a similar virus. (This is one of the reasons why flu vaccine effectiveness is so hit-and-miss: the vaccine must be reformulated every year, based on an informed guess about which strains are likely to cause most suffering in the coming year.) Another type of change is an abrupt "shift" in genetic make-up, caused by a random mutation. When this happens, the possibility of a pandemic occurs: most people will possess no immune protection against the new virus. In short, if the microbial world is our enemy then we face a crafty foe. We might well choose to think of microbes as a threat to our civilisation.

On the other hand, our understanding of medicine in general and of public health in particular are vastly more advanced than in London's day. Furthermore, information can travel even more quickly than people. These advances help mitigate the threat of pandemic disease. Consider, for example, the case of SARS. Between November 2002 and July 2003, a viral disease causing flu-like symptoms caused 774 deaths in China and neighbouring countries. The disease, which was given the name Severe Acute Respiratory Syndrome (SARS for short) was new and it was dangerous—it had a 9.6% fatality rate. There was no vaccine against SARS; there isn't one now. Nevertheless, the dire threat posed to our civilisation by SARS did not materialise. The rapid response of public health authorities, at national and international level, broke the chain of transmission. Since 2004, no case of SARS has been reported anywhere in the world.

Or consider the case of the swine flu outbreak of 2009. A new strain of the H1N1 virus (see Fig. 3.2) began circulating and, since H1N1 caused the dreadful Spanish flu pandemic, it's no surprise that individuals and organisations were worried. Fortunately, the virus that caused swine flu was about one hundred times less lethal than the 1918 virus. Even if that had not been the case, I suspect the outcome of the 2009 pandemic would have been less severe than what happened in 1918. For one thing, doctors could prescribe antiviral medicines—the antivirals weren't hugely successful, but they were better than nothing. More importantly, the public health response was rapid. My own university was soon covered with posters explaining how to slow the transmission of the disease. Some of those posters are still to be found; they are fading, now, but they still provide basic but effective hygiene advice. Furthermore, although it turned out not to be needed, organisations developed business continuity plans. In my own university, these continuity plans involved teaching online if a pandemic caused students to stay away from lecture theatres. (In 1665, the University of Cambridge closed down because of pandemic. In this case it was a precaution against the Great Plague, the last major outbreak of bubonic plague in England. Cambridge was unable to offer online learning back in 1665, but that turned out not to be a hindrance for Isaac Newton. During enforced private study at his home in Woolsthorpe he developed optics, calculus, and the law of universal gravitation!)

I'm writing this almost exactly one hundred years after the doctors observed the first cases of Spanish flu. For an entire century, humans have managed to avoid a widespread outbreak of contagious disease. So it's tempting to conclude that although technological developments might promote a pandemic

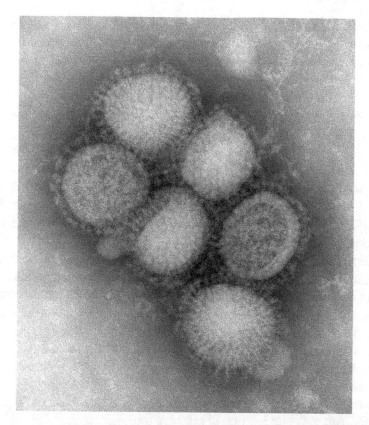

Fig. 3.2 A transmission electron micrograph (see Chapter 11) of the H1N1 influenza virus, which was responsible for the 2009 swine flu outbreak. An H1N1 virus caused the much more deadly Spanish flu outbreak of 1918 (Credit: C.S. Goldsmith and A. Balish, Centers for Disease Control and Prevention)

they also provide us with tools to help prevent a pandemic. It's comforting to suppose that our science, technology, and medicine will avert the disaster envisaged in "The Scarlet Plague".

Our knowledge might well save us. But only a fool would be complacent.

* * *

Humanity faces a growing threat from bacteria—a self-inflicted threat we could ward off if people would only act rationally. The menace stems from the misuse of our most effective weapon against bacteria.

People are naturally enamoured of antibiotics. Patients demand, and often receive, antibiotics as a treatment for colds, sore throats, earaches… even in cases where doctors know antibiotics won't work. Antibiotics are used in crop production, as pesticides, or to treat disease in plants. Antibiotics are given to animals as freely as they are given to humans—vets use them to treat disease while farmers use them to promote growth. We live in a world awash with antibiotics. The problem? Well, in order to survive in this antibiotic-filled environment bacteria have evolved resistance. Some bacteria are now resistant to all known antibiotics.

A world without effective antibiotics is a terrifying prospect. Many routine medical interventions we now take for granted—appendix operations, hip replacements, transplant surgery—would be dangerous: patients might survive the knife but succumb to infection. We'd face the same risks people faced before 1928, when Alexander Fleming discovered penicillin. Worse, though, is that possibility of pandemic.

Consider the plague. The disease is caused by the bacterium *Yersinia pestis*. When this pestilence stalked our ancestors the prognosis was poor for anyone infected: chances were high the sufferer would die a horrible death. Antibiotics changed the story. The plague infects people to this day, but nowadays if patients are given streptomycin quickly enough then the chances are high they'll survive. So if a strain of *Yersinia pestis* evolves resistance to all antibiotics then doctors—and, more to the point, patients—will be in trouble. It's a similar story with many other infectious diseases.

Science and medicine might have provided us with tools to help prevent a pandemic, but we're letting one of our best tools get rusty.

*   *   *

Black death, Spanish flu, or something along the lines of the scarlet plague—another pandemic will surely happen eventually. When it comes, though, we'll at least have the comfort of knowing the pandemic agent won't be *trying* to kill us. Some bacteria and viruses cause us harm, but that's just a byproduct of their life cycle. It's nothing personal. As mentioned in Chapter 2, however, advances in biotechnology will soon permit a group or even an individual to *design* a microbial agent that's *intended* to kill. The terrorist, or perhaps merely the misogynist, will have the ability to target disease at specific groups—males, females, the pre-pubescent, those possessing too much or too little skin melanin. And an engineered pandemic could be designed to kill more effectively than the natural variety.

Consider the West African Ebola epidemic of 2013–2016. According to official statistics the virus caused 11,310 deaths. This was a shocking outbreak, of course, but in some ways the virus killed too effectively for its own good. Symptoms became obvious between 2 and 21 days after exposure to the virus, and this led to a method of containment and control. Anyone who was in contact with a patient was tracked for 21 days; communities were made aware of risk factors and preventative measures; and quarantines were put in place. Roughly 900 days after the first case was diagnosed, the epidemic was over. But the single-minded bioterrorist could *engineer* an infectious agent in such a way that obvious countermeasures would be ineffective. The agent, for example, might be airborne and easily transmitted through the simple act of breathing. It might establish itself in its host whilst producing no symptoms. After a long latency period it could be "switched on"—and death would follow for all those infected.

SF writers have long imagined something along these lines. In a 1982 novel, for example, Frank Herbert's titular "white plague" was designed by a molecular biologist. The biologist, driven insane by the death of his wife and children in a car bomb, desires revenge—so he develops a deadly plague, which is carried by men but kills only women. In his 1997 novel *The Cobra Event*, Richard Preston has an antagonist called Archimedes release a genetically engineered virus (called "Cobra")—a fusion of the common cold and smallpox viruses—which results in a horrifying disease called brainpox. (Preston is the author of the non-fiction book *The Hot Zone*, which gives a well written account of the viral haemorrhagic fevers.) In Paulo Bacigalupi's award-winning 2009 novel *The Windup Girl*, large corporations release bioengineered plagues that attack crops—a lucrative activity if you possess plague-resistant seeds.

White plague, Cobra, bioengineered attack genes... these nightmares remain science fictional. The knowledge and techniques needed to realise them don't exist yet. But, as described in the commentary to the previous chapter, the rate of progress in biotechnology is astounding. In a few years there'll be research labs possessing the knowledge necessary to create a deadly life form from scratch; a few years later those same techniques will be available to undergraduates.

Such technology is so dangerous it presents an existential risk. Should we not therefore police it, in the same way we police other existential risks? After all, the world has managed to avoid a nuclear catastrophe by cooperating at the international level to limit the spread of the technology that permits the construction of hydrogen bombs. Unfortunately, nuclear non-proliferation techniques won't work in the case of bioterrorism. The construction of a nuclear arsenal can't easily be hidden from view—the resources of a nation state are required to build a hydrogen bomb—and so the activity can in principle be monitored. But one day soon a small terrorist cell working quietly in a garden shed might be able to engineer a virus using only a few test tubes. How could society possibly police that situation? The human race can survive natural pandemics: there have been many in our past and there'll be more in our future. But could humanity survive an *engineered* pandemic?

* * *

If a pandemic, natural or artificial, did wipe out humanity, what might Earth be like without us? Following London, it's interesting to speculate. Many of our buildings would likely soon vanish under the onslaught of wind, rain, and vegetation; roads would crack and bridges would collapse; a few constructions—the Channel Tunnel, for example—might last much longer. But eventually, most traces of humanity's time on this planet would be erased. Perhaps after a few tens of millions of years—the same sort of timescale separating us from the dinosaurs—evolution might produce another intelligent species. Would all traces of the achievements we so value be dissolved by the passage of time? Or would such a species be able to find evidence that humans once walked the Earth?

* * *

If some future intelligence was indeed able to infer the existence of a bipedal creature, which chose to dig up fossilized carbon and transform it into plastics and a source of power, then we'd surely seem bizarre to them. But would we seem as bizarre as the creatures depicted in Chapter 4?

Notes and Further Reading

the impact was so terrible—The Spanish flu is the subject of numerous books; for an entertaining account (if "entertaining" is the correct word to use about such a grim event), see Spinney (2017).

attacked humans throughout history—In *Plagues and Peoples*, William McNeill (1976) explores the effects of disease on human history. He gives an in-depth account of the Justinian Plague, as well as more recent pandemics.

the Black Death in 1347—One of the best of many accounts of the Black Death, its causes and consequences, is that by John Kelly (2006).

According to official statistics—For further details of the Ebola epidemic that peaked during 2014–2015, see World Health Organisation (2016). For more recent information about this terrible disease, see World Health Organisation (2018).

writers have long imagined something along these lines—For the books and novels mentioned here, see: Bacigalupi (2009); Herbert (1982); and Preston (1995, 1997).

be able to find evidence that we once walked the Earth?—This question lies at the core of a fascinating novel by a world-renowned astrophysicist. Jayant Narlikar's (2015) novel *The Return of Vaman* appears in the Springer Science & Fiction series. For non-fiction accounts exploring what the world might be like if humans vanished, see for example Weisman (2007) and Zalasiewicz (2009).

Bibliography

Bacigalupi, P.: The Windup Girl. Night Shade Books, San Francisco (2009)

Herbert, F.: The White Plague. G.P. Putnam's, New York (1982)

Kelly, J.: The Great Mortality: An Intimate History of the Black Death, the Most Devastating Plague of All Time. Harper Perennial, New York (2006)

London, J.: The scarlet plague. London Magazine. June issue (1912)

McNeill, W.H.: Plagues and Peoples. Anchor, New York (1976)

Narlikar, J.V.: The Return of Vaman: A Scientific Novel. Springer, Berlin (2015)

Preston, R.: The Hot Zone. Anchor, New York (1995)

Preston, R.: The Cobra Event. Random House, New York (1997)

Spiney, L.: Pale Rider: The Spanish Flu of 1918 and How It Changed the World. Public Affairs, New York (2017)

Weisman, A.: The World Without Us. St Martin's Press, New York (2007)

World Health Organisation: Ebola outbreak 2014–15. http://www.who.int/csr/disease/ebola/en/ (2016)

World Health Organisation: Ebola virus disease. http://www.who.int/ebola/en/ (2018)

Zalasiewicz, J.: The Earth After Us: What Legacy Will Humans Leave in the Rocks? OUP, Oxford (2009)

4

Life on Mars

A Martian Odyssey (Stanley G. Weinbaum)

Jarvis stretched himself as luxuriously as he could in the cramped general quarters of the *Ares*.

"Air you can breathe!" he exulted. "It feels as thick as soup after the thin stuff out there!" He nodded at the Martian landscape stretching flat and desolate in the light of the nearer moon, beyond the glass of the port.

The other three stared at him sympathetically—Putz, the engineer, Leroy, the biologist, and Harrison, the astronomer and captain of the expedition. Dick Jarvis was chemist of the famous crew, the *Ares* expedition, first human beings to set foot on the mysterious neighbor of the earth, the planet Mars. This, of course, was in the old days, less than twenty years after the mad American Doheny perfected the atomic blast at the cost of his life, and only a decade after the equally mad Cardoza rode on it to the moon. They were true pioneers, these four of the *Ares*. Except for a half-dozen moon expeditions and the ill-fated de Lancey flight aimed at the seductive orb of Venus, they were the first men to feel other gravity than earth's, and certainly the first successful crew to leave the earth–moon system. And they deserved that success when one considers the difficulties and discomforts—the months spent in acclimatization chambers back on earth, learning to breathe the air as tenuous as that of Mars, the challenging of the void in the tiny rocket driven by the cranky reaction motors of the twenty-first century, and mostly the facing of an absolutely unknown world.

© Springer Nature Switzerland AG 2019
S. Webb, *New Light Through Old Windows: Exploring Contemporary Science Through 12 Classic Science Fiction Tales*, Science and Fiction,
https://doi.org/10.1007/978-3-030-03195-4_4

Jarvis stretched and fingered the raw and peeling tip of his frost-bitten nose. He sighed again contentedly.

"Well," exploded Harrison abruptly, "are we going to hear what happened? You set out all shipshape in an auxiliary rocket, we don't get a peep for ten days, and finally Putz here picks you out of a lunatic ant-heap with a freak ostrich as your pal! Spill it, man!"

"Speel?" queried Leroy perplexedly. "Speel what?"

"He means 'spiel'," explained Putz soberly. "It iss to tell."

Jarvis met Harrison's amused glance without the shadow of a smile. "That's right, Karl," he said in grave agreement with Putz. "*Ich spiel es!*" He grunted comfortably and began.

"According to orders," he said, "I watched Karl here take off toward the North, and then I got into my flying sweat-box and headed South. You'll remember, Cap—we had orders not to land, but just scout about for points of interest. I set the two cameras clicking and buzzed along, riding pretty high— about two thousand feet—for a couple of reasons. First, it gave the cameras a greater field, and second, the under-jets travel so far in this half-vacuum they call air here that they stir up dust if you move low."

"We know all that from Putz," grunted Harrison. "I wish you'd saved the films, though. They'd have paid the cost of this junket; remember how the public mobbed the first moon pictures?"

"The films are safe," retorted Jarvis. "Well," he resumed, "as I said, I buzzed along at a pretty good clip; just as we figured, the wings haven't much lift in this air at less than a hundred miles per hour, and even then I had to use the under-jets.

"So, with the speed and the altitude and the blurring caused by the under-jets, the seeing wasn't any too good. I could see enough, though, to distinguish that what I sailed over was just more of this grey plain that we'd been examining the whole week since our landing—same blobby growths and the same eternal carpet of crawling little plant-animals, or biopods, as Leroy calls them. So I sailed along, calling back my position every hour as instructed, and not knowing whether you heard me."

"I did!" snapped Harrison.

"A hundred and fifty miles south," continued Jarvis imperturbably, "the surface changed to a sort of low plateau, nothing but desert and orange-tinted sand. I figured that we were right in our guess, then, and this grey plain we dropped on was really the Mare Cimmerium which would make my orange desert the region called Xanthus. If I were right, I ought to hit another grey plain, the Mare Chronium in another couple of hundred miles, and then another orange desert, Thyle I or II. And so I did."

"Putz verified our position a week and a half ago!" grumbled the captain. "Let's get to the point."

"Coming!" remarked Jarvis. "Twenty miles into Thyle—believe it or not—I crossed a canal!"

"Putz photographed a hundred! Let's hear something new!"

"And did he also see a city?"

"Twenty of 'em, if you call those heaps of mud cities!"

"Well," observed Jarvis, "from here on I'll be telling a few things Putz didn't see!" He rubbed his tingling nose, and continued. "I knew that I had sixteen hours of daylight at this season, so eight hours—eight hundred miles—from here, I decided to turn back. I was still over Thyle, whether I or II I'm not sure, not more than twenty-five miles into it. And right there, Putz's pet motor quit!"

"Quit? How?" Putz was solicitous.

"The atomic blast got weak. I started losing altitude right away, and suddenly there I was with a thump right in the middle of Thyle! Smashed my nose on the window, too!" He rubbed the injured member ruefully.

"Did you maybe try vashing der combustion chamber mit acid sulphuric?" inquired Putz. "Sometimes der lead giffs a secondary radiation—"

"Naw!" said Jarvis disgustedly. "I wouldn't try that, of course—not more than ten times! Besides, the bump flattened the landing gear and busted off the under-jets. Suppose I got the thing working—what then? Ten miles with the blast coming right out of the bottom and I'd have melted the floor from under me!" He rubbed his nose again. "Lucky for me a pound only weighs seven ounces here, or I'd have been mashed flat!"

"I could have fixed!" ejaculated the engineer. "I bet it vas not serious."

"Probably not," agreed Jarvis sarcastically. "Only it wouldn't fly. Nothing serious, but I had my choice of waiting to be picked up or trying to walk back—eight hundred miles, and perhaps twenty days before we had to leave! Forty miles a day! Well," he concluded, "I chose to walk. Just as much chance of being picked up, and it kept me busy."

"We'd have found you," said Harrison.

"No doubt. Anyway, I rigged up a harness from some seat straps, and put the water tank on my back, took a cartridge belt and revolver, and some iron rations, and started out."

"Water tank!" exclaimed the little biologist, Leroy. "She weigh one-quarter ton!"

"Wasn't full. Weighed about two hundred and fifty pounds earth-weight, which is eighty-five here. Then, besides, my own personal two hundred and ten pounds is only seventy on Mars, so, tank and all, I grossed a hundred and fifty-five, or fifty-five pounds less than my everyday earth-weight. I figured on that when I undertook the forty-mile daily stroll. Oh—of course I took a thermo-skin sleeping bag for these wintry Martian nights.

"Off I went, bouncing along pretty quickly. Eight hours of daylight meant twenty miles or more. It got tiresome, of course—plugging along over a soft sand desert with nothing to see, not even Leroy's crawling biopods. But an hour or so brought me to the canal—just a dry ditch about four hundred feet wide, and straight as a railroad on its own company map.

"There'd been water in it sometime, though. The ditch was covered with what looked like a nice green lawn. Only, as I approached, the lawn moved out of my way!"

"Eh?" said Leroy.

"Yeah, it was a relative of your biopods. I caught one—a little grass-like blade about as long as my finger, with two thin, stemmy legs."

"He is where?" Leroy was eager.

"He is let go! I had to move, so I plowed along with the walking grass opening in front and closing behind. And then I was out on the orange desert of Thyle again.

"I plugged steadily along, cussing the sand that made going so tiresome, and, incidentally, cussing that cranky motor of yours, Karl. It was just before twilight that I reached the edge of Thyle, and looked down over the gray Mare Chronium. And I knew there was seventy-five miles of *that* to be walked over, and then a couple of hundred miles of that Xanthus desert, and about as much more Mare Cimmerium. Was I pleased? I started cussing you fellows for not picking me up!"

"We were trying, you sap!" said Harrison.

"That didn't help. Well, I figured I might as well use what was left of daylight in getting down the cliff that bounded Thyle. I found an easy place, and down I went. Mare Chronium was just the same sort of place as this—crazy leafless plants and a bunch of crawlers; I gave it a glance and hauled out my sleeping bag. Up to that time, you know, I hadn't seen anything worth worrying about on this half-dead world—nothing dangerous, that is."

"Did you?" queried Harrison.

"*Did I!* You'll hear about it when I come to it. Well, I was just about to turn in when suddenly I heard the wildest sort of shenanigans!"

"Vot iss shenanigans?" inquired Putz.

"He says, 'Je ne sais quoi,'" explained Leroy. "It is to say, 'I don't know what.'"

"That's right," agreed Jarvis. "I didn't know what, so I sneaked over to find out. There was a racket like a flock of crows eating a bunch of canaries—whistles, cackles, caws, trills, and what have you. I rounded a clump of stumps, and there was Tweel!"

"Tweel?" said Harrison, and "Tveel?" said Leroy and Putz.

"That freak ostrich," explained the narrator. "At least, Tweel is as near as I can pronounce it without sputtering. He called it something like 'Trrrweerrlll.'"

"What was he doing?" asked the Captain.

"He was being eaten! And squealing, of course, as any one would."

"Eaten! By what?"

"I found out later. All I could see then was a bunch of black ropy arms tangled around what looked like, as Putz described it to you, an ostrich. I wasn't going to interfere, naturally; if both creatures were dangerous, I'd have one less to worry about.

"But the bird-like thing was putting up a good battle, dealing vicious blows with an eighteen-inch beak, between screeches. And besides, I caught a glimpse or two of what was on the end of those arms!" Jarvis shuddered. "But the clincher was when I noticed a little black bag or case hung about the neck of the bird-thing! It was intelligent! That or tame, I assumed. Anyway, it clinched my decision. I pulled out my automatic and fired into what I could see of its antagonist.

"There was a flurry of tentacles and a spurt of black corruption, and then the thing, with a disgusting sucking noise, pulled itself and its arms into a hole in the ground. The other let out a series of clacks, staggered around on legs about as thick as golf sticks, and turned suddenly to face me. I held my weapon ready, and the two of us stared at each other.

"The Martian wasn't a bird, really. It wasn't even bird-like, except just at first glance. It had a beak all right, and a few feathery appendages, but the beak wasn't really a beak. It was somewhat flexible; I could see the tip bend slowly from side to side; it was almost like a cross between a beak and a trunk. It had four-toed feet, and four fingered things—hands, you'd have to call them, and a little roundish body, and a long neck ending in a tiny head—and that beak. It stood an inch or so taller than I, and—well, Putz saw it!"

The engineer nodded. "*Ja!* I saw!"

Jarvis continued. "So—we stared at each other. Finally the creature went into a series of clackings and twitterings and held out its hands toward me, empty. I took that as a gesture of friendship."

"Perhaps," suggested Harrison, "it looked at that nose of yours and thought you were its brother!"

"Huh! You can be funny without talking! Anyway, I put up my gun and said 'Aw, don't mention it,' or something of the sort, and the thing came over and we were pals.

"By that time, the sun was pretty low and I knew that I'd better build a fire or get into my thermo-skin. I decided on the fire. I picked a spot at the base of the Thyle cliff, where the rock could reflect a little heat on my back. I started breaking off chunks of this desiccated Martian vegetation, and my companion caught the idea and brought in an armful. I reached for a match, but the Martian fished into his pouch and brought out something that looked like a glowing coal; one touch of it, and the fire was blazing—and you all know what a job we have starting a fire in this atmosphere!

"And that bag of his!" continued the narrator. "That was a manufactured article, my friends; press an end and she popped open—press the middle and she sealed so perfectly you couldn't see the line. Better than zippers.

"Well, we stared at the fire a while and I decided to attempt some sort of communication with the Martian. I pointed at myself and said 'Dick'; he caught the drift immediately, stretched a bony claw at me and repeated 'Tick.' Then I pointed at him, and he gave that whistle I called Tweel; I can't imitate his accent. Things were going smoothly; to emphasize the names, I repeated 'Dick,' and then, pointing at him, 'Tweel.'

"There we stuck! He gave some clacks that sounded negative, and said something like 'P-p-p-proot.' And that was just the beginning; I was always 'Tick,' but as for him—part of the time he was 'Tweel,' and part of the time he was 'P-p-p-proot,' and part of the time he was sixteen other noises!

"We just couldn't connect. I tried 'rock,' and I tried 'star,' and 'tree,' and 'fire,' and Lord knows what else, and try as I would, I couldn't get a single word! Nothing was the same for two successive minutes, and if that's a language, I'm an alchemist! Finally I gave it up and called him Tweel, and that seemed to do.

"But Tweel hung on to some of my words. He remembered a couple of them, which I suppose is a great achievement if you're used to a language you have to make up as you go along. But I couldn't get the hang of his talk; either I missed some subtle point or we just didn't think alike—and I rather believe the latter view.

"I've other reasons for believing that. After a while I gave up the language business, and tried mathematics. I scratched two plus two equals four on the ground, and demonstrated it with pebbles. Again Tweel caught the idea, and informed me that three plus three equals six. Once more we seemed to be getting somewhere.

"So, knowing that Tweel had at least a grammar school education, I drew a circle for the sun, pointing first at it, and then at the last glow of the sun. Then I sketched in Mercury, and Venus, and Mother Earth, and Mars, and finally, pointing to Mars, I swept my hand around in a sort of inclusive gesture to indicate that Mars was our current environment. I was working up to putting over the idea that my home was on the earth.

"Tweel understood my diagram all right. He poked his beak at it, and with a great deal of trilling and clucking, he added Deimos and Phobos to Mars, and then sketched in the earth's moon!

"Do you see what that proves? It proves that Tweel's race uses telescopes—that they're civilized!"

"Does not!" snapped Harrison. "The moon is visible from here as a fifth magnitude star. They could see its revolution with the naked eye."

"The moon, yes!" said Jarvis. "You've missed my point. Mercury isn't visible! And Tweel knew of Mercury because he placed the moon at the *third* planet, not the second. If he didn't know Mercury, he'd put the earth second, and Mars third, instead of fourth! See?"

"Humph!" said Harrison.

"Anyway," proceeded Jarvis, "I went on with my lesson. Things were going smoothly, and it looked as if I could put the idea over. I pointed at the earth on my diagram, and then at myself, and then, to clinch it, I pointed to myself and then to the earth itself shining bright green almost at the zenith.

"Tweel set up such an excited clacking that I was certain he understood. He jumped up and down, and suddenly he pointed at himself and then at the sky, and then at himself and at the sky again. He pointed at his middle and then at Arcturus, at his head and then at Spica, at his feet and then at half a dozen stars, while I just gaped at him. Then, all of a sudden, he gave a tremendous leap. Man, what a hop! He shot straight up into the starlight, seventy-five feet if an inch! I saw him silhouetted against the sky, saw him turn and come down at me head first, and land smack on his beak like a javelin! There he stuck square in the center of my sun-circle in the sand—a bull's eye!"

"Nuts!" observed the captain. "Plain nuts!"

"That's what I thought, too! I just stared at him open-mouthed while he pulled his head out of the sand and stood up. Then I figured he'd missed my point, and I went through the whole blamed rigamarole again, and it ended the same way, with Tweel on his nose in the middle of my picture!"

"Maybe it's a religious rite," suggested Harrison.

"Maybe," said Jarvis dubiously. "Well, there we were. We could exchange ideas up to a certain point, and then—blooey! Something in us was different, unrelated; I don't doubt that Tweel thought me just as screwy as I thought him. Our minds simply looked at the world from different viewpoints, and perhaps his viewpoint is as true as ours. But—we couldn't get together, that's all. Yet, in spite of all difficulties, I *liked* Tweel, and I have a queer certainty that he liked me."

"Nuts!" repeated the captain. "Just daffy!"

"Yeah? Wait and see. A couple of times I've thought that perhaps we—" He paused, and then resumed his narrative. "Anyway, I finally gave it up, and got into my thermo-skin to sleep. The fire hadn't kept me any too warm, but that damned sleeping bag did. Got stuffy five minutes after I closed myself in. I opened it a little and bingo! Some eighty-below-zero air hit my nose, and that's when I got this pleasant little frostbite to add to the bump I acquired during the crash of my rocket.

"I don't know what Tweel made of my sleeping. He sat around, but when I woke up, he was gone. I'd just crawled out of my bag, though, when I heard some twittering, and there he came, sailing down from that three-story Thyle cliff to alight on his beak beside me. I pointed to myself and toward the north, and he pointed at himself and toward the south, but when I loaded up and started away, he came along.

"Man, how he traveled! A hundred and fifty feet at a jump, sailing through the air stretched out like a spear, and landing on his beak. He seemed surprised at my plodding, but after a few moments he fell in beside me, only every few minutes he'd go into one of his leaps, and stick his nose into the sand a block ahead of me. Then he'd come shooting back at me; it made me nervous at first to see that beak of his coming at me like a spear, but he always ended in the sand at my side.

"So the two of us plugged along across the Mare Chronium. Same sort of place as this—same crazy plants and same little green biopods growing in the sand, or crawling out of your way. We talked—not that we understood each other, you know, but just for company. I sang songs, and I suspect Tweel did too; at least, some of his trillings and twitterings had a subtle sort of rhythm.

"Then, for variety, Tweel would display his smattering of English words. He'd point to an outcropping and say 'rock,' and point to a pebble and say it again; or he'd touch my arm and say 'Tick,' and then repeat it. He seemed terrifically amused that the same word meant the same thing twice in succession, or that the same word could apply to two different objects. It set me wondering if perhaps his language wasn't like the primitive speech of some earth people—you know, Captain, like the Negritoes, for instance, who haven't any generic words. No word for food or water or man—words for good food and bad food, or rain water and sea water, or strong man and weak man—but no names for general classes. They're too primitive to understand that rain water and sea water are just different aspects of the same thing. But that wasn't the case with Tweel; it was just that we were somehow mysteriously different—our minds were alien to each other. And yet—we *liked* each other!"

"Looney, that's all," remarked Harrison. "That's why you two were so fond of each other."

"Well, I like *you!*" countered Jarvis wickedly. "Anyway," he resumed, "don't get the idea that there was anything screwy about Tweel. In fact, I'm not so sure but that he couldn't teach our highly praised human intelligence a trick or two. Oh, he wasn't an intellectual superman, I guess; but don't overlook the point that he managed to understand a little of my mental workings, and I never even got a glimmering of his."

"Because he didn't have any!" suggested the captain, while Putz and Leroy blinked attentively.

"You can judge of that when I'm through," said Jarvis. "Well, we plugged along across the Mare Chronium all that day, and all the next. Mare Chronium—Sea of Time! Say, I was willing to agree with Schiaparelli's name by the end of that march! Just that grey, endless plain of weird plants, and never a sign of any other life. It was so monotonous that I was even glad to see the desert of Xanthus toward the evening of the second day.

"I was fair worn out, but Tweel seemed as fresh as ever, for all I never saw him drink or eat. I think he could have crossed the Mare Chronium in a couple of hours with those block-long nose dives of his, but he stuck along with me. I offered him some water once or twice; he took the cup from me and sucked the liquid into his beak, and then carefully squirted it all back into the cup and gravely returned it.

"Just as we sighted Xanthus, or the cliffs that bounded it, one of those nasty sand clouds blew along, not as bad as the one we had here, but mean to travel against. I pulled the transparent flap of my thermo-skin bag across my face and managed pretty well, and I noticed that Tweel used some feathery append-ages growing like a mustache at the base of his beak to cover his nostrils, and some similar fuzz to shield his eyes."

"He is a desert creature!" ejaculated the little biologist, Leroy. "Huh? Why?"

"He drink no water—he is adapt' for sand storm—"

"Proves nothing! There's not enough water to waste anywhere on this desic-cated pill called Mars. We'd call all of it desert on earth, you know." He paused. "Anyway, after the sand storm blew over, a little wind kept blowing in our faces, not strong enough to stir the sand. But suddenly things came drifting along from the Xanthus cliffs—small, transparent spheres, for all the world like glass tennis balls! But light—they were almost light enough to float even in this thin air—empty, too; at least, I cracked open a couple and nothing came out but a bad smell. I asked Tweel about them, but all he said was 'No, no, no,' which I took to mean that he knew nothing about them. So they went bouncing by like tumbleweeds, or like soap bubbles, and we plugged on toward Xanthus. Tweel pointed at one of the crystal balls once and said 'rock,' but I was too tired to argue with him. Later I discovered what he meant.

"We came to the bottom of the Xanthus cliffs finally, when there wasn't much daylight left. I decided to sleep on the plateau if possible; anything dangerous, I reasoned, would be more likely to prowl through the vegetation of the Mare Chronium than the sand of Xanthus. Not that I'd seen a single sign of menace, except the rope-armed black thing that had trapped Tweel, and apparently that didn't prowl at all, but lured its victims within reach. It couldn't lure me while I slept, especially as Tweel didn't seem to sleep at all, but simply sat patiently around all night. I wondered how the creature had managed to trap Tweel, but there wasn't any way of asking him. I found that out too, later; it's devilish!

"However, we were ambling around the base of the Xanthus barrier looking for an easy spot to climb. At least, I was. Tweel could have leaped it easily, for the cliffs were lower than Thyle—perhaps sixty feet. I found a place and started up, swearing at the water tank strapped to my back—it didn't bother me except when climbing—and suddenly I heard a sound that I thought I recognized!

"You know how deceptive sounds are in this thin air. A shot sounds like the pop of a cork. But this sound was the drone of a rocket, and sure enough, there went our second auxiliary about ten miles to westward, between me and the sunset!"

"Vas me!" said Putz. "I hunt for you."

"Yeah; I knew that, but what good did it do me? I hung on to the cliff and yelled and waved with one hand. Tweel saw it too, and set up a trilling and twittering, leaping to the top of the barrier and then high into the air. And while I watched, the machine droned on into the shadows to the south.

"I scrambled to the top of the cliff. Tweel was still pointing and trilling excitedly, shooting up toward the sky and coming down head-on to stick upside down on his beak in the sand. I pointed toward the south and at myself, and he said, 'Yes—Yes—Yes'; but somehow I gathered that he thought the flying thing was a relative of mine, probably a parent. Perhaps I did his intellect an injustice; I think now that I did.

"I was bitterly disappointed by the failure to attract attention. I pulled out my thermo-skin bag and crawled into it, as the night chill was already apparent. Tweel stuck his beak into the sand and drew up his legs and arms and looked for all the world like one of those leafless shrubs out there. I think he stayed that way all night."

"Protective mimicry!" ejaculated Leroy. "See? He is desert creature!"

"In the morning," resumed Jarvis, "we started off again. We hadn't gone a hundred yards into Xanthus when I saw something queer! This is one thing Putz didn't photograph, I'll wager!

"There was a line of little pyramids—tiny ones, not more than six inches high, stretching across Xanthus as far as I could see! Little buildings made of pygmy bricks, they were, hollow inside and truncated, or at least broken at the top and empty. I pointed at them and said 'What?' to Tweel, but he gave some negative twitters to indicate, I suppose, that he didn't know. So off we went, following the row of pyramids because they ran north, and I was going north.

"Man, we trailed that line for hours! After a while, I noticed another queer thing: they were getting larger. Same number of bricks in each one, but the bricks were larger.

"By noon they were shoulder high. I looked into a couple—all just the same, broken at the top and empty. I examined a brick or two as well; they were silica, and old as creation itself!"

"How you know?" asked Leroy.

"They were weathered—edges rounded. Silica doesn't weather easily even on earth, and in this climate—!"

"How old you think?"

"Fifty thousand—a hundred thousand years. How can I tell? The little ones we saw in the morning were older—perhaps ten times as old. Crumbling. How old would that make *them*? Half a million years? Who knows?" Jarvis paused a moment. "Well," he resumed, "we followed the line. Tweel pointed at them and said 'rock' once or twice, but he'd done that many times before. Besides, he was more or less right about these.

"I tried questioning him. I pointed at a pyramid and asked 'People?' and indicated the two of us. He set up a negative sort of clucking and said, 'No, no, no. No one-one-two. No two-two-four,' meanwhile rubbing his stomach. I just stared at him and he went through the business again 'No one-one-two. No two-two-four.' I just gaped at him."

"That proves it!" exclaimed Harrison. "Nuts!"

"You think so?" queried Jarvis sardonically. "Well, I figured it out different! 'No one-one-two!' You don't get it, of course, do you?"

"Nope—nor do you!"

"I think I do! Tweel was using the few English words he knew to put over a very complex idea. What, let me ask, does mathematics make you think of?"

"Why—of astronomy. Or—or logic!"

"That's it! 'No one-one-two!' Tweel was telling me that the builders of the pyramids weren't people—or that they weren't intelligent, that they weren't reasoning creatures! Get it?"

"Huh! I'll be damned!"

"You probably will."

"Why," put in Leroy, "he rub his belly?"

"Why? Because, my dear biologist, that's where his brains are! Not in his tiny head—in his middle!"

"*C'est* impossible!"

"Not on Mars, it isn't! This flora and fauna aren't earthly; your biopods prove that!" Jarvis grinned and took up his narrative. "Anyway, we plugged along across Xanthus and in about the middle of the afternoon, something else queer happened. The pyramids ended."

"Ended!"

"Yeah; the queer part was that the last one—and now they were ten-footers—was capped! See? Whatever built it was still inside; we'd trailed 'em from their half-million-year-old origin to the present.

"Tweel and I noticed it about the same time. I yanked out my automatic (I had a clip of Boland explosive bullets in it) and Tweel, quick as a sleight-of-hand trick, snapped a queer little glass revolver out of his bag. It was much like our weapons, except that the grip was larger to accommodate his four-taloned hand. And we held our weapons ready while we sneaked up along the lines of empty pyramids.

"Tweel saw the movement first. The top tiers of bricks were heaving, shaking, and suddenly slid down the sides with a thin crash. And then—something—something was coming out!

"A long, silvery-grey arm appeared, dragging after it an armored body. Armored, I mean, with scales, silver-grey and dull-shining. The arm heaved the body out of the hole; the beast crashed to the sand.

"It was a nondescript creature—body like a big grey cask, arm and a sort of mouth-hole at one end; stiff, pointed tail at the other—and that's all. No other limbs, no eyes, ears, nose—nothing! The thing dragged itself a few yards, inserted its pointed tail in the sand, pushed itself upright, and just sat.

"Tweel and I watched it for ten minutes before it moved. Then, with a creaking and rustling like—oh, like crumpling stiff paper—its arm moved to the mouth-hole and out came a brick! The arm placed the brick carefully on the ground, and the thing was still again.

"Another ten minutes—another brick. Just one of Nature's bricklayers. I was about to slip away and move on when Tweel pointed at the thing and said 'rock'! I went 'huh?' and he said it again. Then, to the accompaniment of some of his trilling, he said, 'No—no—,' and gave two or three whistling breaths.

"Well, I got his meaning, for a wonder! I said, 'No breath?' and demonstrated the word. Tweel was ecstatic; he said, 'Yes, yes, yes! No, no, no breet!' Then he gave a leap and sailed out to land on his nose about one pace from the monster!

"I was startled, you can imagine! The arm was going up for a brick, and I expected to see Tweel caught and mangled, but—nothing happened! Tweel pounded on the creature, and the arm took the brick and placed it neatly beside the first. Tweel rapped on its body again, and said 'rock,' and I got up nerve enough to take a look myself.

"Tweel was right again. The creature was rock, and it didn't breathe!"

"How you know?" snapped Leroy, his black eyes blazing interest. "Because I'm a chemist. The beast was made of silica! There must have been pure silicon in the sand, and it lived on that. Get it? We, and Tweel, and those plants out there, and even the biopods are *carbon* life; this thing lived by a different set of chemical reactions. It was silicon life!"

"*La vie silicieuse!*" shouted Leroy. "I have suspect, and now it is proof! I must go see! *Il faut que je—*"

"All right! All right!" said Jarvis. "You can go see. Anyhow, there the thing was, alive and yet not alive, moving every ten minutes, and then only to remove a brick. Those bricks were its waste matter. See, Frenchy? We're carbon, and our waste is carbon dioxide, and this thing is silicon, and *its* waste is silicon dioxide—silica. But silica is a solid, hence the bricks. And it builds itself in, and when it is covered, it moves over to a fresh place to start over. No wonder it creaked! A living creature half a million years old!"

"How you know how old?" Leroy was frantic.

"We trailed its pyramids from the beginning, didn't we? If this weren't the original pyramid builder, the series would have ended somewhere before we found him, wouldn't it?—ended and started over with the small ones. That's simple enough, isn't it?

"But he reproduces, or tries to. Before the third brick came out, there was a little rustle and out popped a whole stream of those little crystal balls. They're his spores, or eggs, or seeds—call 'em what you want. They went bouncing by across Xanthus just as they'd bounced by us back in the Mare Chronium. I've a hunch how they work, too—this is for your information, Leroy. I think the crystal shell of silica is no more than a protective covering, like an eggshell, and that the active principle is the smell inside. It's some sort of gas that attacks silicon, and if the shell is broken near a supply of that element, some reaction starts that ultimately develops into a beast like that one."

"You should try!" exclaimed the little Frenchman. "We must break one to see!"

"Yeah? Well, I did. I smashed a couple against the sand. Would you like to come back in about ten thousand years to see if I planted some pyramid monsters? You'd most likely be able to tell by that time!" Jarvis paused and drew a deep breath. "Lord! That queer creature! Do you picture it? Blind, deaf, nerveless, brainless—just a mechanism, and yet—immortal! Bound to go on making bricks, building pyramids, as long as silicon and oxygen exist, and even afterwards it'll just stop. It won't be dead. If the accidents of a million years bring it its food again, there it'll be, ready to run again, while brains and civilizations are part of the past. A queer beast—yet I met a stranger one!"

"If you did, it must have been in your dreams!" growled Harrison. "You're right!" said Jarvis soberly. "In a way, you're right. The dream-beast! That's the best name for it—and it's the most fiendish, terrifying creation one could imagine! More dangerous than a lion, more insidious than a snake!"

"Tell me!" begged Leroy. "I must go see!"

"Not *this* devil!" He paused again. "Well," he resumed, "Tweel and I left the pyramid creature and plowed along through Xanthus. I was tired and a little disheartened by Putz's failure to pick me up, and Tweel's trilling got on my nerves, as did his flying nosedives. So I just strode along without a word, hour after hour across that monotonous desert.

"Toward mid-afternoon we came in sight of a low dark line on the horizon. I knew what it was. It was a canal; I'd crossed it in the rocket and it meant that we were just one-third of the way across Xanthus. Pleasant thought, wasn't it? And still, I was keeping up to schedule.

"We approached the canal slowly; I remembered that this one was bordered by a wide fringe of vegetation and that Mud-heap City was on it.

"I was tired, as I said. I kept thinking of a good hot meal, and then from that I jumped to reflections of how nice and home-like even Borneo would seem after this crazy planet, and from that, to thoughts of little old New York, and then to thinking about a girl I know there—Fancy Long. Know her?"

"Vision entertainer," said Harrison. "I've tuned her in. Nice blonde—dances and sings on the *Yerba Mate* hour."

"That's her," said Jarvis ungrammatically. "I know her pretty well—just friends, get me?—though she came down to see us off in the *Ares*. Well, I was thinking about her, feeling pretty lonesome, and all the time we were approaching that line of rubbery plants.

"And then—I said, 'What 'n Hell!' and stared. And there she was—Fancy Long, standing plain as day under one of those crack-brained trees, and smiling and waving just the way I remembered her when we left!"

"Now you're nuts, too!" observed the captain.

"Boy, I almost agreed with you! I stared and pinched myself and closed my eyes and then stared again—and every time, there was Fancy Long smiling and waving! Tweel saw something, too; he was trilling and clucking away, but I scarcely heard him. I was bounding toward her over the sand, too amazed even to ask myself questions.

"I wasn't twenty feet from her when Tweel caught me with one of his flying leaps. He grabbed my arm, yelling, 'No—no—no!' in his squeaky voice. I tried to shake him off—he was as light as if he were built of bamboo—but he dug his claws in and yelled. And finally some sort of sanity returned to me and I stopped less than ten feet from her. There she stood, looking as solid as Putz's head!"

"Vot?" said the engineer.

"She smiled and waved, and waved and smiled, and I stood there dumb as Leroy, while Tweel squeaked and chattered. I *knew* it couldn't be real, yet—there she was!

"Finally I said, 'Fancy! Fancy Long!' She just kept on smiling and waving, but looking as real as if I hadn't left her thirty-seven million miles away.

"Tweel had his glass pistol out, pointing it at her. I grabbed his arm, but he tried to push me away. He pointed at her and said, 'No breet! No breet!' and I understood that he meant that the Fancy Long thing wasn't alive. Man, my head was whirling!

"Still, it gave me the jitters to see him pointing his weapon at her. I don't know why I stood there watching him take careful aim, but I did. Then he squeezed the handle of his weapon; there was a little puff of steam, and Fancy Long was gone! And in her place was one of those writhing, black, rope-armed horrors like the one I'd saved Tweel from!

"The dream-beast! I stood there dizzy, watching it die while Tweel trilled and whistled. Finally he touched my arm, pointed at the twisting thing, and said, 'You one-one-two, he one-one-two.' After he'd repeated it eight or ten times, I got it. Do any of you?"

"*Oui!*" shrilled Leroy. "*Moi—je le comprends!* He mean you think of something, the beast he know, and you see it! *Un chien*—a hungry dog, he would see the big bone with meat! Or smell it—not?"

"Right!" said Jarvis. "The dream-beast uses its victim's longings and desires to trap its prey. The bird at nesting season would see its mate, the fox, prowling for its own prey, would see a helpless rabbit!"

"How he do?" queried Leroy.

"How do I know? How does a snake back on earth charm a bird into its very jaws? And aren't there deep-sea fish that lure their victims into their mouths? Lord!" Jarvis shuddered. "Do you see how insidious the monster is? We're warned now—but henceforth we can't trust even our eyes. You might see me—I might see one of you—and back of it may be nothing but another of those black horrors!"

"How'd your friend know?" asked the captain abruptly.

"Tweel? I wonder! Perhaps he was thinking of something that couldn't possibly have interested me, and when I started to run, he realized that I saw something different and was warned. Or perhaps the dream-beast can only project a single vision, and Tweel saw what I saw—or nothing. I couldn't ask him. But it's just another proof that his intelligence is equal to ours or greater."

"He's daffy, I tell you!" said Harrison. "What makes you think his intellect ranks with the human?"

"Plenty of things! First, the pyramid-beast. He hadn't seen one before; he said as much. Yet he recognized it as a dead-alive automaton of silicon."

"He could have heard of it," objected Harrison. "He lives around here, you know."

"Well how about the language? I couldn't pick up a single idea of his and he learned six or seven words of mine. And do you realize what complex ideas he put over with no more than those six or seven words? The pyramid-monster—the dream-beast! In a single phrase he told me that one was a harmless automaton and the other a deadly hypnotist. What about that?"

"Huh!" said the captain.

"*Huh* if you wish! Could you have done it knowing only six words of English? Could you go even further, as Tweel did, and tell me that another creature was of a sort of intelligence so different from ours that understanding was impossible—even more impossible than that between Tweel and me?"

"Eh? What was that?"

"Later. The point I'm making is that Tweel and his race are worthy of our friendship. Somewhere on Mars—and you'll find I'm right—is a civilization and culture equal to ours, and maybe more than equal. And communication is possible between them and us; Tweel proves that. It may take years of patient trial, for their minds are alien, but less alien than the next minds we encountered—if they *are* minds."

"The next ones? What next ones?"

"The people of the mud cities along the canals." Jarvis frowned, then resumed his narrative. "I thought the dream-beast and the silicon-monster were the strangest beings conceivable, but I was wrong. These creatures are still more alien, less understandable than either and far less comprehensible than Tweel, with whom friendship is possible, and even, by patience and concentration, the exchange of ideas.

"Well," he continued, "we left the dream-beast dying, dragging itself back into its hole, and we moved toward the canal. There was a carpet of that queer walking-grass scampering out of our way, and when we reached the bank, there was a yellow trickle of water flowing. The mound city I'd noticed from the rocket was a mile or so to the right and I was curious enough to want to take a look at it.

"It had seemed deserted from my previous glimpse of it, and if any creatures were lurking in it—well, Tweel and I were both armed. And by the way, that crystal weapon of Tweel's was an interesting device; I took a look at it after the dream-beast episode. It fired a little glass splinter, poisoned, I suppose, and I guess it held at least a hundred of 'em to a load. The propellent was steam—just plain steam!"

"Shteam!" echoed Putz. "From vot come, shteam?"

"From water, of course! You could see the water through the transparent handle and about a gill of another liquid, thick and yellowish. When Tweel squeezed the handle—there was no trigger—a drop of water and a drop of the yellow stuff squirted into the firing chamber, and the water vaporized—pop!—like that. It's not so difficult; I think we could develop the same principle. Concentrated sulphuric acid will heat water almost to boiling, and so will quicklime, and there's potassium and sodium—

"Of course, his weapon hadn't the range of mine, but it wasn't so bad in this thin air, and it *did* hold as many shots as a cowboy's gun in a Western movie. It was effective, too, at least against Martian life; I tried it out, aiming at one of the crazy plants, and darned if the plant didn't wither up and fall apart! That's why I think the glass splinters were poisoned.

"Anyway, we trudged along toward the mud-heap city and I began to wonder whether the city builders dug the canals. I pointed to the city and then at the canal, and Tweel said 'No—no—no!' and gestured toward the south. I took it to mean that some other race had created the canal system, perhaps Tweel's people. I don't know; maybe there's still another intelligent race on the planet, or a dozen others. Mars is a queer little world.

"A hundred yards from the city we crossed a sort of road—just a hard-packed mud trail, and then, all of a sudden, along came one of the mound builders!

"Man, talk about fantastic beings! It looked rather like a barrel trotting along on four legs with four other arms or tentacles. It had no head, just body and members and a row of eyes completely around it. The top end of the barrel-body was a diaphragm stretched as tight as a drum head, and that was all. It was pushing a little coppery cart and tore right past us like the proverbial bat out of Hell. It didn't even notice us, although I thought the eyes on my side shifted a little as it passed.

"A moment later another came along, pushing another empty cart. Same thing—it just scooted past us. Well, I wasn't going to be ignored by a bunch of barrels playing train, so when the third one approached, I planted myself in the way—ready to jump, of course, if the thing didn't stop.

"But it did. It stopped and set up a sort of drumming from the diaphragm on top. And I held out both hands and said, 'We are friends!' And what do you suppose the thing did?"

"Said, 'Pleased to meet you,' I'll bet!" suggested Harrison.

"I couldn't have been more surprised if it had! It drummed on its diaphragm, and then suddenly boomed out, 'We are v-r-r-riends!' and gave its pushcart a vicious poke at me! I jumped aside, and away it went while I stared dumbly after it.

"A minute later another one came hurrying along. This one didn't pause, but simply drummed out, 'We are v-r-r-riends!' and scurried by. How did it learn the phrase? Were all of the creatures in some sort of communication with each other? Were they all parts of some central organism? I don't know, though I think Tweel does.

"Anyway, the creatures went sailing past us, every one greeting us with the same statement. It got to be funny; I never thought to find so many friends on this God-forsaken ball! Finally I made a puzzled gesture to Tweel; I guess he understood, for he said, 'One-one-two—yes!—two-two-four—no!' Get it?"

"Sure," said Harrison, "It's a Martian nursery rhyme."

"Yeah! Well, I was getting used to Tweel's symbolism, and I figured it out this way. 'One-one-two—yes!' The creatures were intelligent. 'Two-two-four—no!' Their intelligence was not of our order, but something different and beyond the logic of two and two is four. Maybe I missed his meaning. Perhaps he meant that their minds were of low degree, able to figure out the simple things—'One-one-two—yes!'—but not more difficult things—'Two-two-four—no!' But I think from what we saw later that he meant the other.

"After a few moments, the creatures came rushing back—first one, then another. Their pushcarts were full of stones, sand, chunks of rubbery plants, and such rubbish as that. They droned out their friendly greeting, which didn't really sound so friendly, and dashed on. The third one I assumed to be my first acquaintance and I decided to have another chat with him. I stepped into his path again and waited.

"Up he came, booming out his 'We are v-r-r-riends' and stopped. I looked at him; four or five of his eyes looked at me. He tried his password again and gave a shove on his cart, but I stood firm. And then the—the dashed creature reached out one of his arms, and two finger-like nippers tweaked my nose!"

"Haw!" roared Harrison. "Maybe the things have a sense of beauty!"

"Laugh!" grumbled Jarvis. "I'd already had a nasty bump and a mean frost-bite on that nose. Anyway, I yelled 'Ouch!' and jumped aside and the creature dashed away; but from then on, their greeting was 'We are v-r-r-riends! Ouch!' Queer beasts!

"Tweel and I followed the road squarely up to the nearest mound. The creatures were coming and going, paying us not the slightest attention, fetching their loads of rubbish. The road simply dived into an opening, and slanted down like an old mine, and in and out darted the barrel-people, greeting us with their eternal phrase.

"I looked in; there was a light somewhere below, and I was curious to see it. It didn't look like a flame or torch, you understand, but more like a civilized light, and I thought that I might get some clue as to the creatures' development. So in I went and Tweel tagged along, not without a few trills and twitters, however.

"The light was curious; it sputtered and flared like an old arc light, but came from a single black rod set in the wall of the corridor. It was electric, beyond doubt. The creatures were fairly civilized, apparently.

"Then I saw another light shining on something that glittered and I went on to look at that, but it was only a heap of shiny sand. I turned toward the entrance to leave, and the Devil take me if it wasn't gone!

"I suppose the corridor had curved, or I'd stepped into a side passage. Anyway, I walked back in that direction I thought we'd come, and all I saw was more dimlit corridor. The place was a labyrinth! There was nothing but twisting passages running every way, lit by occasional lights, and now and then a creature running by, sometimes with a pushcart, sometimes without. "Well, I wasn't much worried at first. Tweel and I had only come a few steps from the entrance. But every move we made after that seemed to get us in deeper. Finally I tried following one of the creatures with an empty cart, thinking that he'd be going out for his rubbish, but he ran around aimlessly, into one passage and out another. When he started dashing around a pillar like one of these Japanese waltzing mice, I gave up, dumped my water tank on the floor, and sat down.

"Tweel was as lost as I. I pointed up and he said 'No—no—no!' in a sort of helpless trill. And we couldn't get any help from the natives. They paid no attention at all, except to assure us they were friends—ouch!

"Lord! I don't know how many hours or days we wandered around there! I slept twice from sheer exhaustion; Tweel never seemed to need sleep. We tried following only the upward corridors, but they'd run uphill a ways and then curve downwards. The temperature in that damned ant hill was constant; you couldn't tell night from day and after my first sleep I didn't know whether I'd slept one hour or thirteen, so I couldn't tell from my watch whether it was midnight or noon.

"We saw plenty of strange things. There were machines running in some of the corridors, but they didn't seem to be doing anything—just wheels turning. And several times I saw two barrel-beasts with a little one growing between them, joined to both."

"Parthenogenesis!" exulted Leroy. "Parthenogenesis by budding like *les tulipes*!"

"If you say so, Frenchy," agreed Jarvis. "The things never noticed us at all, except, as I say, to greet us with 'We are v-r-r-riends! Ouch!' They seemed to have no home-life of any sort, but just scurried around with their pushcarts, bringing in rubbish. And finally I discovered what they did with it.

"We'd had a little luck with a corridor, one that slanted upwards for a great distance. I was feeling that we ought to be close to the surface when suddenly the passage debouched into a domed chamber, the only one we'd seen. And man!—I felt like dancing when I saw what looked like daylight through a crevice in the roof.

"There was a—a sort of machine in the chamber, just an enormous wheel that turned slowly, and one of the creatures was in the act of dumping his rubbish below it. The wheel ground it with a crunch—sand, stones, plants, all into powder that sifted away somewhere. While we watched, others filed in, repeating the process, and that seemed to be all. No rhyme nor reason to the whole thing—but that's characteristic of this crazy planet. And there was another fact that's almost too bizarre to believe.

"One of the creatures, having dumped his load, pushed his cart aside with a crash and calmly shoved himself under the wheel! I watched him being crushed, too stupefied to make a sound, and a moment later, another followed him! They were perfectly methodical about it, too; one of the cartless creatures took the abandoned pushcart.

"Tweel didn't seem surprised; I pointed out the next suicide to him, and he just gave the most human-like shrug imaginable, as much as to say, 'What can I do about it?' He must have known more or less about these creatures.

"Then I saw something else. There was something beyond the wheel, something shining on a sort of low pedestal. I walked over; there was a little crystal about the size of an egg, fluorescing to beat Tophet. The light from it stung my hands and face, almost like a static discharge, and then I noticed another funny thing. Remember that wart I had on my left thumb? Look!" Jarvis extended his hand. "It dried up and fell off—just like that! And my abused nose—say, the pain went out of it like magic! The thing had the property of hard x-rays or gamma radiations, only more so; it destroyed diseased tissue and left healthy tissue unharmed!

"I was thinking what a present *that'd* be to take back to Mother Earth when a lot of racket interrupted. We dashed back to the other side of the wheel in time to see one of the pushcarts ground up. Some suicide had been careless, it seems.

"Then suddenly the creatures were booming and drumming all around us and their noise was decidedly menacing. A crowd of them advanced toward us; we backed out of what I thought was the passage we'd entered by, and they came rumbling after us, some pushing carts and some not. Crazy brutes! There was a whole chorus of 'We are v-r-r-riends! Ouch!' I didn't like the 'ouch'; it was rather suggestive.

"Tweel had his glass gun out and I dumped my water tank for greater freedom and got mine. We backed up the corridor with the barrel-beasts following—about twenty of them. Queer thing—the ones coming in with loaded carts moved past us inches away without a sign.

"Tweel must have noticed that. Suddenly, he snatched out that glowing coal cigar-lighter of his and touched a cart-load of plant limbs. Puff! The whole load was burning—and the crazy beast pushing it went right along without a change of pace! It created some disturbance among our 'V-r-r-riends,' however—and then I noticed the smoke eddying and swirling past us, and sure enough, there was the entrance!

"I grabbed Tweel and out we dashed and after us our twenty pursuers. The daylight felt like Heaven, though I saw at first glance that the sun was all but set, and that was bad, since I couldn't live outside my thermo-skin bag in a Martian night—at least, without a fire.

"And things got worse in a hurry. They cornered us in an angle between two mounds, and there we stood. I hadn't fired nor had Tweel; there wasn't any use in irritating the brutes. They stopped a little distance away and began their booming about friendship and ouches.

"Then things got still worse! A barrel-brute came out with a pushcart and they all grabbed into it and came out with handfuls of foot-long copper darts—sharp-looking ones—and all of a sudden one sailed past my ear—zing! And it was shoot or die then.

"We were doing pretty well for a while. We picked off the ones next to the pushcart and managed to keep the darts at a minimum, but suddenly there was a thunderous booming of 'v-r-r-riends' and 'ouches,' and a whole army of 'em came out of their hole.

"Man! We were through and I knew it! Then I realized that Tweel wasn't. He could have leaped the mound behind us as easily as not. He was staying for me!

"Say, I could have cried if there'd been time! I'd liked Tweel from the first, but whether I'd have had gratitude to do what he was doing—suppose I *had* saved him from the first dream-beast—he'd done as much for me, hadn't he? I grabbed his arm, and said 'Tweel,' and pointed up, and he understood. He said, 'No—no—no, Tick!' and popped away with his glass pistol.

"What could I do? I'd be a goner anyway when the sun set, but I couldn't explain that to him. I said, 'Thanks, Tweel. You're a man!' and felt that I wasn't paying him any compliment at all. A man! There are mighty few men who'd do that.

"So I went 'bang' with my gun and Tweel went 'puff ' with his, and the barrels were throwing darts and getting ready to rush us, and booming about being friends. I had given up hope. Then suddenly an angel dropped right down from Heaven in the shape of Putz, with his under-jets blasting the barrels into very small pieces!

"Wow! I let out a yell and dashed for the rocket; Putz opened the door and in I went, laughing and crying and shouting! It was a moment or so before I remembered Tweel; I looked around in time to see him rising in one of his nosedives over the mound and away.

"I had a devil of a job arguing Putz into following! By the time we got the rocket aloft, darkness was down; you know how it comes here—like turning off a light. We sailed out over the desert and put down once or twice. I yelled 'Tweel!' and yelled it a hundred times, I guess. We couldn't find him; he could travel like the wind and all I got—or else I imagined it—was a faint trilling and twittering drifting out of the south. He'd gone, and damn it! I wish—I wish he hadn't!"

The four men of the *Ares* were silent—even the sardonic Harrison. At last little Leroy broke the stillness.

"I should like to see," he murmured.

"Yeah," said Harrison. "And the wart-cure. Too bad you missed that; it might be the cancer cure they've been hunting for a century and a half."

"Oh, that!" muttered Jarvis gloomily. "That's what started the fight!" He drew a glistening object from his pocket.

"Here it is."

Stanley Grauman Weinbaum (1902–1935) published his first story, "A Martian Odyssey", in July 1934. Just 17 months later, he died from lung cancer. Weinbaum's SF career may have been short, but his influence on the field was great. In particular, his groundbreaking "A Martian Odyssey" represented a new kind of SF story, a type of story championed by John W Campbell: one in which aliens are sympathetic but nevertheless non-human, able to think as *well* as a human but not *like* a human.

Commentary

Each year the Science Fiction Writers of America (SFWA) present Nebula Awards for the best SF novel, novella, novelette, and short story published in the preceding twelve months. The Nebula Award is a relatively recent invention (well it's younger than I am, which in my mind makes it of recent vintage) so, in order to honour stories that predated the first Nebula Awards, the SFWA organized a poll to decide on the greatest SF published before 1965. The winner of the best short science fiction story generated no surprise: it was "Nightfall", by Isaac Asimov. That "A Martian Odyssey" finished runner-up was perhaps more surprising—Weinbaum's writing is, to put it kindly, of the standard one would expect from a story appearing in the July 1934 issue of a pulp magazine such as *Wonder Stories*. But the accolade was in truth well deserved. Asimov himself, in his introduction to *The Best of Stanley G. Weinbaum,* wrote that this tale of Martian adventure "had the effect on the field of an exploding grenade. With this single story Weinbaum was instantly recognized as the world's best living science fiction writer, and at once almost every writer in the field tried to imitate him." Weinbaum didn't get the opportunity to be prolific; the cancer that killed him was likely establishing itself just as he was imagining the remarkable properties of the cart creatures' cancer-healing crystal. But this single story of first contact with aliens sealed his fame. It remains one of the most important stories in the history of science fiction.

Although the story's dialogue and plotting are dated, and its setting is unconvincing, Weinbaum's vision of extraterrestrial life remains fresh. The story succeeds because of the sheer invention Weinbaum brings to his description of alien life forms. Indeed, there's more of biological interest on his fictional Mars than there is here on Earth. As well as carbon-based flora and fauna there are organisms based on silicon—and those ancient, silicon-based pyramid-builders are wonderfully enigmatic. The dream beast is a fascinating and compelling creature—and even if its method of luring prey might not be physically possible it possesses a certain evolutionary logic. The cart creatures are immensely intriguing because they simply don't care about humans—they have their own, unknowable motivations. And Tweel—well, Tweel is one of the great aliens in all of science fiction. The Mars of Weinbaum's imagination would be a tremendous place for biologists to explore. Unfortunately, even before Weinbaum published his story, astronomers knew Mars was unlikely to be home to advanced life forms.

The nineteenth century had seen many speculations about Mars and the possibility of life. Telescopic observations had shown that Mars has ice caps, like Earth, which grow and shrink throughout the year; that Mars has seasons, like Earth; and that the Martian day is roughly the same length as Earth's day. So in 1877, when Giovanni Schiaparelli observed patterns on the surface of the Red Planet and called them "canali", many people were primed to think of Martian "canals" (the product of technology) rather than "channels" (the product of nature). In 1894, Percival Lowell set up the Flagstaff Observatory and spent much of the rest of his life investigating and popularizing the notion that an advanced Martian civilisation had built canals to transfer water from the ice caps to the rest of a drying, dying planet. Unfortunately for Lowell and his grand vision, in the same year he founded his observatory an American astronomer, William Wallace Campbell, discovered the Martian atmosphere contains neither water nor oxygen in anything other than trace amounts. This discovery didn't *prove* life was absent on Mars—but it really didn't boost the idea. Even worse, no one else could reliably see the markings that Lowell saw. And in 1909 astronomers trained the world's largest telescope on Mars—and they saw not canals, or even channels, but irregular geological features. Those surface markings that Lowell painstakingly drew were just optical illusions. Nothing exists on Mars that resembles Tweel, the dream beast, or the pyramid-builders.

So "A Martian Odyssey" is not and never could have been a field guide to Martian zoology. But might the Red Planet be home to *simple* life? And even if it turns out the planet is barren *now*, might there have been simple life early in its history? These are interesting questions, which science is now on the verge of answering.

*   *   *

At the time of writing, space agencies have made 44 attempts to send craft—flyby missions and orbiters, landers and rovers—to Mars. (Of these missions, 21 have been US attempts; 18 have been USSR/Russian; Europe, India, and

Japan have each sent one mission; and there have been joint European/Russian and Chinese/Russian missions.) One can gauge the difficulty of space exploration by noting that 23 of these attempts ended in failure and a further three can only be classed as partial successes. Nevertheless, despite the immense technical hurdles engineers must overcome, 18 missions have succeeded in studying Mars both at a distance (with orbiting satellites) and close-up (with surface rovers). The results are intriguing.

In January 2004, two Mars Exploration Rovers—*Spirit* and *Opportunity*—landed and began what was planned to be a 90-day research mission. *Spirit* continued its explorations until 2009, when it got trapped in soft soil and was no longer able to move; *Opportunity*, at the time of writing, continues to explore the Red Planet. The intriguing result is that both *Spirit* and *Opportunity* have found signs that liquid water once flowed on Mars. Furthermore the *Curiosity* Rover, which touched down in 2012 with a mission to explore the Gale crater, quickly found clear evidence for a fast-running stream that, aeons ago, flowed knee-deep through the region. The conclusion: a few billion years ago Mars was warm and wet. And where there's water there's at least the chance of life. When the solar system was still young might Mars have possessed an environment in which life could have flourished? (Fig. 4.1).

To be clear: there is no evidence that any organism lives on Mars now. Mars and Earth followed different evolutionary paths. Since Mars is smaller than Earth it cooled more quickly, and its liquid iron–nickel core froze solid. Its magnetic field switched off, and with it vanished a protective shield that safeguards a planetary atmosphere from cosmic rays. Over time, as the Sun aged and brightened, Mars was stripped of its atmosphere. The Martian surface turned to rust. *Spirit*, *Opportunity*, and *Curiosity* are not encountering liquid water now. But billions of years ago there *could* have been liquid water flowing on Mars; the temperatures *could* have been rather pleasant; life—simple, microbial life—*could* have survived there, for a while at least. And if microorganisms did establish themselves on Mars then perhaps, even after all these billions of years, we might be able to find some traces that life left behind?

Some astrobiologists already claim to have found evidence for fossilised microbial Martian life—fossils found not on Mars but here on Earth! Allan Hills 84001, a 1.93 kg meteorite uncovered in Antarctica in 1984, was ejected from the Martian surface via meteor impact about 17 million years ago. It landed on Earth about 13,000 years ago. But the rock itself is old—it probably crystallised on Mars about 4 billion years ago, at a time when liquid water was abundant. In 1996, NASA scientists used a scanning electron microscope to study the rock. They saw structures—which they interpreted as being the fossils of incredibly tiny, bacteria-like organisms. Life! This interpretation has

Fig. 4.1 A photograph taken by the Curiosity rover of a rock outcrop in the Gale crater. The outcrop (named Hottah) is exposed bedrock, which consists of small fragments that have been cemented together. Some of these fragments are rounded pebbles, the nature of which causes scientists to theorize that water once flowed across Gale crater. And where water flows there's a chance of life! Even if there was once life on Mars, however, it surely would not have been as abundant as the flora and fauna that appears in Weinbaum's "A Martian Odyssey"! (Credit: NASA/JPL-Caltech/MSSS)

not been widely accepted by the scientific community. There are undoubtedly structures in AH84001 but they can be explained as crystallisations of organic molecules and minerals. We'd need much better evidence than this if we wanted to persuade ourselves we were looking at the remains of extraterrestrial life forms. Nevertheless, the fact such an interpretation is even possible surely means it's worth contemplating what sorts of traces a primitive Martian biosphere might have left behind.

Consider, for example, *Spirit*'s discovery of the mineral opaline silica in the Gusev crater. The formation resembles cauliflower florets sprinkled over the Martian surface (see Fig. 4.2 top). More recently, geologists found similar-looking formations near geysers at the edge of Chile's Atacama desert (see Fig. 4.2 bottom). Importantly, the environment at Atacama is about as close as you can come here on Earth to the Martian environment: the air is thin and dry, the Sun beats down with high levels of UV light, and the night is bitterly

Fig. 4.2 (Top) Mineral formations photographed by the Spirit rover in 2008, in the so-called Home Plate plateau in the Gusev crater on Mars. (Bottom) Hydrothermal mineral deposits near the El Tatio hotsprings in Atacama, Chile. The nodules here grew thanks to the activity of microorganisms. Visually, the cauliflower-like nodules in Atacama look similar to the nodules seen on Mars; could microorganisms have helped cause the Martian structures? (Credit: top—NASA/JPL-Caltech; bottom—Steve Ruff)

cold. So if it turns out the Atacama formations were created with the help of microbial life, as seems plausible, then perhaps microbes caused the Gusev formations? Of course, pointing out similarities proves nothing. To test the idea that microbial life caused the Gusev formations we'd need to analyse Martian samples in detail—but *Spirit* can't help because it's now stuck, presumed dead. Perhaps the next Martian rover will return to Gusev and investigate further. Even better would be if a craft were to travel to Mars, collect samples, and then return them to laboratories here on Earth for analysis. I'm sure one day this will happen.

Within the solar system Mars is not the only place we might look for life. Titan, Saturn's largest moon, is the site of complex chemistry that might conceivably harbour life. A more plausible candidate is Enceladus, another moon of Saturn, which potentially has lakes of liquid water beneath its frozen surface. And astrobiologists have postulated even more exotic places—the upper reaches of the Venusian atmosphere, the buried oceans of Jovian satellites—as being places to search for biology. But Mars remains our best bet for finding extraterrestrial life. Not *extant* life, necessarily, but at least signs of past life. Back when the solar system was young, Earth and Mars were alike; if life started here, it might have started there as well. Who knows—perhaps Mars was where Earth life started: it could have been transferred to a lifeless Earth by meteor.

After a period of investigation, of course, we might be forced to conclude that our neighbour is barren and always has been barren. That even though wet and warm conditions were once in place, life just didn't start on Mars. Perhaps the first life on Mars will be us—when humans set foot there. And what if we learn that Earth, of all the places in the solar system, is the only home for life? Well, perhaps it hints that life—and in particular intelligent, self-conscious life—is rare. It would be wonderful to live in a universe where extraterrestrial life is as common and as strange as Weinbaum's inventions, a universe in which we might try to communicate with non-human intelligences such as Tweel. But maybe the universe contains just us?

*   *   *

There is another possibility for non-human intelligence: perhaps humans, in their pursuit of ever more effective digital technologies, will develop *artificial* intelligence. This thought brings us nicely to Chapter 5.

Notes and Further Reading

Asimov himself, in his introduction—See Weinbaum (1974).
story of first contact with aliens—The question of "first contact" with alien beings is one of the key themes in SF. The story that gave the theme its name is "First contact" by Murray Leinster (1945), but Weinbaum's story predates Leinster's classic by almost a decade. Knight (1971) is an anthology of ten classic first-contact stories, including Leinster's story, but it's just one of many. Innumerable SF novels explore the theme of first contact, and

it's a popular idea in film and TV too: *Star Trek*, for example, wrestled with the problems and introduced the notion of a "Prime Directive"—the idea that space-faring civilisations should wherever possible avoid contact with technologically less advanced civilisations, because the latter might not be psychologically or culturally equipped to handle the shock of such an encounter.

organisms based on silicon—As far as we know, the basic structural and metabolic functions of all living organisms on Earth are based on carbon compounds. Does it necessarily follow that the structural and metabolic functions of all living organisms *everywhere in the universe* are based on carbon? There are two good reasons why it might. First, carbon is relatively abundant; it is the fourth most abundant element in our galaxy. Second, when it comes to chemistry, carbon is extremely versatile: it can form single, double and triple bonds, along with chains, branched chains, and rings … it provides life with rich opportunities to try out many different types of complex compound. On the other hand, there are elements that resemble carbon. In particular, silicon sits directly below carbon in the periodic table of the elements—so it isn't too outrageous to postulate an alternative biochemistry based on silicon. Indeed, the first such suggestion was made as long ago as 1891, by the German astronomer Julius Scheiner. The idea proved popular, and soon the great SF author H.G. Wells (1894), writing in a non-fiction article, remarked of the hypothesis that "one is startled towards fantastic imaginings by such a suggestion: visions of silicon–aluminium organisms …". To the best of my knowledge, however, Weinbaum's "A Martian Odyssey" is the first appearance in SF of the notion of silicon-based life.

Although it might demonstrate appalling carbon chauvinism, I believe most biologists would agree that silicon is unlikely to be the basis of an alternative biochemistry. Silicon, for one thing, just doesn't have the same versatility as carbon. Perhaps our mastery of silicon chip technology will advance to the stage where we must grant that our robots are "alive"; in that case we might reasonably talk of silicon-based life. But it seems unlikely that a living silicon-based organism will develop through natural processes.

speculations about Mars and the possibility of life—Sheehan (1996) is my favourite account of the evolution of astronomers' understanding of the Red Planet. For an entertaining, short account of Lowell's interest in Mars see Zahnle (2001).

space agencies have made 44 attempts—For a list of the 44 missions to Mars, along with their fate, see NASA (n.d.). This website gives details of a 45th mission—the Mars InSight Lander—which, at the time of writing, is on its way to the Red Planet but still has 5 months' travel before it reaches its destination.

two Mars Exploration Rovers—Squyres (2006) gives a first-person account of a NASA scientist's work on the Mars Exploration Rover mission. Pyle (2014) is a science writer's view of the *Curiosity* mission, while Manning and Simon (2017) gives the story of *Curiosity* from inside NASA—Rob Manning was the mission's Chief Engineer. These books are highly recommended for those who want to learn more of the difficulties involved in exploring Mars, and how scientists and engineers have overcome them.

Some astrobiologists already claim to have found evidence—For the fascinating story of meteorite AH84001, and the scientific, political, and media debate it inspired, see Sawyer (2006).

geologists found similar-looking formations—For further details, see Ruff and Farmer (2016).

Bibliography

Knight, D.: First Contact. Pinnacle, New York (1971)

Leinster, M.: First Contact. Astounding. May issue (1945)

Manning, R., Simon, W.L.: Mars Rover Curiosity: An Inside Account from Curiosity's Chief Engineer. Smithsonian Institute, New York (2017)

NASA: Program & Missions. Historical Log. https://mars.nasa.gov/programmissions/missions/log/ (n.d.)

Pyle, R.: Curiosity: An Inside Look at the Mars Rover Mission and the People Who Made It Happen. Prometheus, New York (2014)

Ruff, S.W., Farmer, J.D.: Silica deposits on Mars with features resembling hot spring biosignatures at El Tatio in Chile. Nat. Commun. 7, 13554 (2016)

Sawyer, K.: The Rock from Mars: A True Detective Story on Two Planets. Random House, New York (2006)

Sheehan, W.: The Planet Mars: A History of Observation and Discovery. University of Arizona Press, Tucson, AZ (1996)

Squyres, S.: Roving Mars: Spirit, Opportunity, and the Exploration of the Red Planet. Hachette, New York (2006)

Weinbaum, S.G.: A Martian Odyssey. Wonder Stories. July issue (1934)

Weinbaum, S.G.: The Best of Stanley G. Weinbaum. Introduction by Isaac Asimov. Ballantine, New York (1974)

Wells, H.G.: Another basis for life. Saturday Review, 22 December, p. 676 (1894)

Zahnle, K.: Decline and fall of the Martian empire. Nature. **412**, 209–213 (2001)

5

Artificial Intelligence

Moxon's Master (Ambrose Bierce)

"Are you serious?—do you really believe that a machine thinks?"

I got no immediate reply; Moxon was apparently intent upon the coals in the grate, touching them deftly here and there with the fire-poker till they signified a sense of his attention by a brighter glow. For several weeks I had been observing in him a growing habit of delay in answering even the most trivial of commonplace questions. His air, however, was that of preoccupation rather than deliberation: one might have said that he had "something on his mind."

Presently he said:

"What is a 'machine'? The word has been variously defined. Here is one definition from a popular dictionary: 'Any instrument or organization by which power is applied and made effective, or a desired effect produced.' Well, then, is not a man a machine? And you will admit that he thinks—or thinks he thinks."

"If you do not wish to answer my question," I said, rather testily, "why not say so?—all that you say is mere evasion. You know well enough that when I say 'machine' I do not mean a man, but something that man has made and controls."

"When it does not control him," he said, rising abruptly and looking out of a window, whence nothing was visible in the blackness of a stormy night. A moment later he turned about and with a smile said: "I beg your pardon;

© Springer Nature Switzerland AG 2019

S. Webb, *New Light Through Old Windows: Exploring Contemporary Science Through 12 Classic Science Fiction Tales*, Science and Fiction,

https://doi.org/10.1007/978-3-030-03195-4_5

I had no thought of evasion. I considered the dictionary man's unconscious testimony suggestive and worth something in the discussion. I can give your question a direct answer easily enough: I do believe that a machine thinks about the work that it is doing."

That was direct enough, certainly. It was not altogether pleasing, for it tended to confirm a sad suspicion that Moxon's devotion to study and work in his machine-shop had not been good for him. I knew, for one thing, that he suffered from insomnia, and that is no light affliction. Had it affected his mind? His reply to my question seemed to me then evidence that it had; perhaps I should think differently about it now. I was younger then, and among the blessings that are not denied to youth is ignorance. Incited by that great stimulant to controversy, I said:

"And what, pray, does it think with—in the absence of a brain?"

The reply, coming with less than his customary delay, took his favorite form of counter-interrogation:

"With what does a plant think—in the absence of a brain?"

"Ah, plants also belong to the philosopher class! I should be pleased to know some of their conclusions; you may omit the premises."

"Perhaps," he replied, apparently unaffected by my foolish irony, "you may be able to infer their convictions from their acts. I will spare you the familiar examples of the sensitive mimosa, the several insectivorous flowers and those whose stamens bend down and shake their pollen upon the entering bee in order that he may fertilize their distant mates. But observe this. In an open spot in my garden I planted a climbing vine. When it was barely above the surface I set a stake into the soil a yard away. The vine at once made for it, but as it was about to reach it after several days I removed it a few feet. The vine at once altered its course, making an acute angle, and again made for the stake. This manoeuvre was repeated several times, but finally, as if discouraged, the vine abandoned the pursuit and ignoring further attempts to divert it traveled to a small tree, further away, which it climbed.

"Roots of the eucalyptus will prolong themselves incredibly in search of moisture. A well-known horticulturist relates that one entered an old drain pipe and followed it until it came to a break, where a section of the pipe had been removed to make way for a stone wall that had been built across its course. The root left the drain and followed the wall until it found an opening where a stone had fallen out. It crept through and following the other side of the wall back to the drain, entered the unexplored part and resumed its journey."

"And all this?"

"Can you miss the significance of it? It shows the consciousness of plants. It proves that they think."

"Even if it did—what then? We were speaking, not of plants, but of machines. They may be composed partly of wood—wood that has no longer vitality—or wholly of metal. Is thought an attribute also of the mineral kingdom?"

"How else do you explain the phenomena, for example, of crystallization?"

"I do not explain them."

"Because you cannot without affirming what you wish to deny, namely, intelligent cooperation among the constituent elements of the crystals. When soldiers form lines, or hollow squares, you call it reason. When wild geese in flight take the form of a letter V you say instinct. When the homogeneous atoms of a mineral, moving freely in solution, arrange themselves into shapes mathematically perfect, or particles of frozen moisture into the symmetrical and beautiful forms of snowflakes, you have nothing to say. You have not even invented a name to conceal your heroic unreason."

Moxon was speaking with unusual animation and earnestness. As he paused I heard in an adjoining room known to me as his "machine-shop," which no one but himself was permitted to enter, a singular thumping sound, as of some one pounding upon a table with an open hand. Moxon heard it at the same moment and, visibly agitated, rose and hurriedly passed into the room whence it came. I thought it odd that anyone else should be in there, and my interest in my friend—with doubtless a touch of unwarrantable curiosity—led me to listen intently, though, I am happy to say, not at the keyhole. There were confused sounds, as of a struggle or scuffle; the floor shook. I distinctly heard hard breathing and a hoarse whisper which said "Damn you!" Then all was silent, and presently Moxon reappeared and said, with a rather sorry smile:

"Pardon me for leaving you so abruptly. I have a machine in there that lost its temper and cut up rough."

Fixing my eyes steadily upon his left cheek, which was traversed by four parallel excoriations showing blood, I said:

"How would it do to trim its nails?"

I could have spared myself the jest; he gave it no attention, but seated himself in the chair that he had left and resumed the interrupted monologue as if nothing had occurred:

"Doubtless you do not hold with those (I need not name them to a man of your reading) who have taught that all matter is sentient, that every atom is a living, feeling, conscious being. I do. There is no such thing as dead, inert matter: it is all alive; all instinct with force, actual and potential; all sensitive to the same forces in its environment and susceptible to the contagion of higher and subtler ones residing in such superior organisms as it may be brought into relation with, as those of man when he is fashioning it into an instrument of his will. It absorbs something of his intelligence and purpose—more of them in proportion to the complexity of the resulting machine and that of its work.

"Do you happen to recall Herbert Spencer's definition of 'life'? I read it thirty years ago. He may have altered it afterward, for anything I know, but in all that time I have been unable to think of a single word that could profitably be changed or added or removed. It seems to me not only the best definition, but the only possible one.

"'Life,' he says, 'is a definite combination of heterogeneous changes, both simultaneous and successive, in correspondence with external coexistences and sequences.'"

"That defines the phenomenon," I said, "but gives no hint of its cause."

"That," he replied, "is all that any definition can do. As Mill points out, we know nothing of cause except as an antecedent—nothing of effect except as a consequent. Of certain phenomena, one never occurs without another, which is dissimilar: the first in point of time we call cause, the second, effect. One who had many times seen a rabbit pursued by a dog, and had never seen rabbits and dogs otherwise, would think the rabbit the cause of the dog.

"But I fear," he added, laughing naturally enough, "that my rabbit is leading me a long way from the track of my legitimate quarry: I'm indulging in the pleasure of the chase for its own sake. What I want you to observe is that in Herbert Spencer's definition of 'life' the activity of a machine is included—there is nothing in the definition that is not applicable to it. According to this sharpest of observers and deepest of thinkers, if a man during his period of activity is alive, so is a machine when in operation. As an inventor and constructor of machines I know that to be true."

Moxon was silent for a long time, gazing absently into the fire. It was growing late and I thought it time to be going, but somehow I did not like the notion of leaving him in that isolated house, all alone except for the presence of some person of whose nature my conjectures could go no further than that it was unfriendly, perhaps malign. Leaning toward him and looking earnestly into his eyes while making a motion with my hand through the door of his workshop, I said:

"Moxon, whom have you in there?"

Somewhat to my surprise he laughed lightly and answered without hesitation:

"Nobody; the incident that you have in mind was caused by my folly in leaving a machine in action with nothing to act upon, while I undertook the interminable task of enlightening your understanding. Do you happen to know that Consciousness is the creature of Rhythm?"

"O bother them both!" I replied, rising and laying hold of my overcoat. "I'm going to wish you good night; and I'll add the hope that the machine which you inadvertently left in action will have her gloves on the next time you think it needful to stop her."

Without waiting to observe the effect of my shot I left the house.

Rain was falling, and the darkness was intense. In the sky beyond the crest of a hill toward which I groped my way along precarious plank sidewalks and across miry, unpaved streets I could see the faint glow of the city's lights, but behind me nothing was visible but a single window of Moxon's house. It glowed with what seemed to me a mysterious and fateful meaning. I knew it was an uncurtained aperture in my friend's "machine-shop," and I had little doubt that he had resumed the studies interrupted by his duties as my instructor in mechanical consciousness and the fatherhood of Rhythm. Odd, and in some degree humorous, as his convictions seemed to me at that time, I could not wholly divest myself of the feeling that they had some tragic relation to his life and character—perhaps to his destiny—although I no longer entertained the notion that they were the vagaries of a disordered mind. Whatever might be thought of his views, his exposition of them was too logical for that. Over and over, his last words came back to me: "Consciousness is the creature of Rhythm." Bald and terse as the statement was, I now found it infinitely alluring. At each recurrence it broadened in meaning and deepened in suggestion. Why, here, (I thought) is something upon which to found a philosophy. If consciousness is the product of rhythm all things are conscious, for all have motion, and all motion is rhythmic. I wondered if Moxon knew the significance and breadth of his thought—the scope of this momentous generalization; or had he arrived at his philosophic faith by the tortuous and uncertain road of observation?

That faith was then new to me, and all Moxon's expounding had failed to make me a convert; but now it seemed as if a great light shone about me, like that which fell upon Saul of Tarsus; and out there in the storm and darkness and solitude I experienced what Lewes calls "The endless variety and excitement of philosophic thought." I exulted in a new sense of knowledge, a new pride of reason. My feet seemed hardly to touch the earth; it was as if I were uplifted and borne through the air by invisible wings.

Yielding to an impulse to seek further light from him whom I now recognized as my master and guide, I had unconsciously turned about, and almost before I was aware of having done so found myself again at Moxon's door. I was drenched with rain, but felt no discomfort. Unable in my excitement to find the doorbell I instinctively tried the knob. It turned and, entering, I mounted the stairs to the room that I had so recently left. All was dark and silent; Moxon, as I had supposed, was in the adjoining room—the "machine-shop." Groping along the wall until I found the communicating door I knocked loudly several times, but got no response, which I attributed to the uproar outside, for the wind was blowing a gale and dashing the rain against the thin walls in sheets. The drumming upon the shingle roof spanning the unceiled room was loud and incessant.

I had never been invited into the machine-shop—had, indeed, been denied admittance, as had all others, with one exception, a skilled metal worker, of whom no one knew anything except that his name was Haley and his habit silence. But in my spiritual exaltation, discretion and civility were alike forgotten and I opened the door. What I saw took all philosophical speculation out of me in short order.

Moxon sat facing me at the farther side of a small table upon which a single candle made all the light that was in the room. Opposite him, his back toward me, sat another person. On the table between the two was a chessboard; the men were playing. I knew little of chess, but as only a few pieces were on the board it was obvious that the game was near its close. Moxon was intensely interested—not so much, it seemed to me, in the game as in his antagonist, upon whom he had fixed so intent a look that, standing though I did directly in the line of his vision, I was altogether unobserved. His face was ghastly white, and his eyes glittered like diamonds. Of his antagonist I had only a back view, but that was sufficient; I should not have cared to see his face.

He was apparently not more than five feet in height, with proportions suggesting those of a gorilla—a tremendous breadth of shoulders, thick, short neck and broad, squat head, which had a tangled growth of black hair and was topped with a crimson fez. A tunic of the same color, belted tightly to the waist, reached the seat—apparently a box—upon which he sat; his legs and feet were not seen. His left forearm appeared to rest in his lap; he moved his pieces with his right hand, which seemed disproportionately long.

I had shrunk back and now stood a little to one side of the doorway and in shadow. If Moxon had looked farther than the face of his opponent he could have observed nothing now, except that the door was open. Something forbade me either to enter or to retire, a feeling—I know not how it came—that I was in the presence of an imminent tragedy and might serve my friend by remaining. With a scarcely conscious rebellion against the indelicacy of the act I remained.

The play was rapid. Moxon hardly glanced at the board before making his moves, and to my unskilled eye seemed to move the piece most convenient to his hand, his motions in doing so being quick, nervous and lacking in precision. The response of his antagonist, while equally prompt in the inception, was made with a slow, uniform, mechanical and, I thought, somewhat theatrical movement of the arm, that was a sore trial to my patience. There was something unearthly about it all, and I caught myself shuddering. But I was wet and cold.

Two or three times after moving a piece the stranger slightly inclined his head, and each time I observed that Moxon shifted his king. All at once the thought came to me that the man was dumb. And then that he was a machine—an automaton chess-player! Then I remembered that Moxon had once spoken to me of having invented such a piece of mechanism, though I did not understand that it had actually been constructed. Was all his talk about the consciousness and intelligence of machines merely a prelude to eventual exhibition of this device—only a trick to intensify the effect of its mechanical action upon me in my ignorance of its secret?

A fine end, this, of all my intellectual transports—my "endless variety and excitement of philosophic thought!" I was about to retire in disgust when something occurred to hold my curiosity. I observed a shrug of the thing's great shoulders, as if it were irritated: and so natural was this—so entirely human—that in my new view of the matter it startled me. Nor was that all, for a moment later it struck the table sharply with its clenched hand. At that gesture Moxon seemed even more startled than I: he pushed his chair a little backward, as in alarm.

Presently Moxon, whose play it was, raised his hand high above the board, pounced upon one of his pieces like a sparrow-hawk and with the exclamation "checkmate!" rose quickly to his feet and stepped behind his chair. The automaton sat motionless.

The wind had now gone down, but I heard, at lessening intervals and progressively louder, the rumble and roll of thunder. In the pauses between I now became conscious of a low humming or buzzing which, like the thunder, grew momentarily louder and more distinct. It seemed to come from the body of the automaton, and was unmistakably a whirring of wheels. It gave me the impression of a disordered mechanism which had escaped the repressive and regulating action of some controlling part—an effect such as might be expected if a pawl should be jostled from the teeth of a ratchet-wheel. But before I had time for much conjecture as to its nature my attention was taken by the strange motions of the automaton itself. A slight but continuous convulsion appeared to have possession of it. In body and head it shook like a man with palsy or an ague chill, and the motion augmented every moment until the entire figure was in violent agitation. Suddenly it sprang to its feet and with a movement almost too quick for the eye to follow shot forward across table and chair, with both arms thrust forth to their full length—the posture and lunge of a diver. Moxon tried to throw himself backward out of reach, but he was too late: I saw the horrible thing's hands close upon his throat, his own clutch its wrists. Then the table was overturned, the candle thrown to the floor and extinguished, and all was black dark. But the noise of the struggle was dreadfully distinct, and most terrible of all were the raucous, squawking sounds made by the strangled man's efforts to breathe. Guided by the infernal hubbub, I sprang to the rescue of my friend, but had hardly taken a stride in the darkness when the whole room blazed with a blinding white light that burned into my brain and heart and memory a vivid picture of the combatants on the floor, Moxon underneath, his throat still in the clutch of those iron hands, his head forced backward, his eyes protruding, his mouth wide open and his tongue thrust out; and—horrible contrast!—upon the painted face of his assassin an expression of tranquil and profound thought, as in the solution of a problem in chess! This I observed, then all was blackness and silence.

Three days later I recovered consciousness in a hospital. As the memory of that tragic night slowly evolved in my ailing brain recognized in my attendant Moxon's confidential workman, Haley. Responding to a look he approached, smiling.

"Tell me about it," I managed to say, faintly—"all about it."

"Certainly," he said; "you were carried unconscious from a burning house— Moxon's. Nobody knows how you came to be there. You may have to do a little explaining. The origin of the fire is a bit mysterious, too. My own notion is that the house was struck by lightning."

"And Moxon?"

"Buried yesterday—what was left of him."

Apparently this reticent person could unfold himself on occasion. When imparting shocking intelligence to the sick he was affable enough. After some moments of the keenest mental suffering I ventured to ask another question:

"Who rescued me?"

"Well, if that interests you—I did."

"Thank you, Mr. Haley, and may God bless you for it. Did you rescue, also, that charming product of your skill, the automaton chess-player that murdered its inventor?"

The man was silent a long time, looking away from me. Presently he turned and gravely said:

"Do you know that?"

"I do," I replied; "I saw it done."

That was many years ago. If asked to-day I should answer less confidently.

Ambrose Gwinnett Bierce (1842–c1914) was an American short story writer, journalist, and critic—the latter career he pursued vigorously while employing the motto 'nothing matters'. His biting and sardonic reviews earned him the nickname 'Bitter Bierce'. At the age of 71 he traveled to Mexico to observe the Mexican Revolution. He disappeared without trace, and to this day the matter of his death remains a mystery.

Commentary

"Moxon's Master" seems straightforward enough—man makes machine, machine murders man—but the famously mordant "Bitter" Bierce often wrote stories where a subtext offers a quite different reading to the surface text, and this story is one of those. SF critics have offered at least three different and diverging interpretations of "Moxon's Master": that Moxon's had a mistress who hid in the automaton and killed Moxon in a fit of rage; that Moxon was gay, and his attempts to seduce the narrator provoked a jealous rage in Haley, who hid in the automaton and killed Moxon; and that the crime was merely staged by Moxon, in order to convince the narrator that machines can be intelligent, and Moxon's death by thunderbolt was the act of a vengeful God. Bierce's tale has also been much commented upon in artificial intelligence (AI) circles. In particular, several AI pioneers argued that the automaton must have been intelligent—and indeed been a conscious, cogitating entity—because it was good at chess. But just as a surface reading of the story might cause us to misunderstand who Moxon's master was, might it have caused those early AI commentators to miss the point of Moxon's chess game?

The rules of chess are, as boardgames go, complicated. And the number of choices faced by chess players during a game—should I move the bishop to that square or this one? or would it be better to move the knight there? or the queen here? or ... —well, the number is immense. This mix of complexity and variety caused most people to believe chess is a game for the "brainy" or intelligent. And perhaps this is why people long believed computers would never beat humans at chess: computers, after all, aren't intelligent. (The Turk, a chess-playing automaton unveiled in 1770 by von Kempelen, caused wonder across Europe and America: the mechanical man was an extremely strong

Fig. 5.1 A reconstruction of von Kempelen's chess-playing Turk, displayed in the Heinz Nixdorf MuseumsForum—the world's biggest museum devoted to computers—in Paderborn, Germany. The Turk was a formidable chess player, but it wasn't an automaton; rather, the Turk was an elaborate trick. A chess master, hidden inside the cabinet, played the game; intricate mechanisms caused the Turk's arms and fingers to move; and von Kempelen provided the sleight-of-hand and diversions needed to distract the audience. Presumably Moxon's automaton would have appeared something like the Turk (Credit: Marcin Wichary, San Francisco)

player, who won the majority of his games. But the Turk was merely a clever hoax. See Fig. 5.1.) Since there isn't a clearly defined winning strategy for chess, as there is for, say, tic-tac-toe, and since the number of possible moves in a game of chess is too large to permit even the most powerful processor to crunch through all the options, people thought computers would only ever be woodpushers—and even then they'd have to get humans to push the wood.

At least, that was the generally accepted belief. AI pioneers thought differently.

As long ago as 1949 Claude Shannon, one of the founders of our modern digital world, showed how one could restrict the number of possibilities a computer would need to consider if it were playing a game of chess. Shannon also showed how a program, if it had been fed the relative values of chess pieces and the values associated with certain positional factors, could evaluate the best move in a given position. In 1950, Alan Turing actually wrote a program to play chess. (Turing's achievement demonstrates that if there's a connection between intelligence and chess playing ability then it's a tenuous one. Turing, by any reasonable definition of the term, was a genius. But he was, apparently, a weak chess player.)

From 1950 onwards, progress in computer chess was slow but relentless. By the 1980s, cheap chess programs were capable of beating the likes of me without breaking sweat. (That was hardly a landmark achievement. I like to think of myself as "brainy", but if I am then my intelligence certainly doesn't extend to chess. The average six-year-old can beat me.) Then, in 1997, the IBM supercomputer Deep Blue beat Garry Kasparov, the then reigning world champion and one of the strongest players of all time. Today, no human can possibly hope to compete in a game against a dedicated chess program running on half-decent hardware. Indeed, technology has advanced to the stage where it's entirely possible for a tech-savvy person to develop a modern-day version of von Kempelen's Turk: an automaton that not only plays an extremely strong game but that moves the pieces for itself. (See Fig. 5.2.)

Proponents of AI argued that Deep Blue's victory was a watershed moment. Dissenters argued (correctly, in my opinion) that it was a mistake to regard chess-playing ability as a signifier of intelligence: just as tic-tac-toe can be defeated by brute force, so can chess. It just takes more force. The dissenters, however, didn't learn their lesson. They began promoting Go, an ancient

Fig. 5.2 Nowadays you can build your own "Turk"—a true automaton that can play chess (at a level that's almost certainly better than your own). As you can see from the photograph, this doesn't look as impressive as the original Turk. Nevertheless, unlike the original Turk, this one needs no human intervention to move chess pieces on the board (Credit: www.raspberryturk.com)

Chinese board game, as an example of an intellectual pursuit that could distinguish humans from machines.

Compared to chess, the rules of Go are simple. But the game of Go contains many more possible moves and so brute-force approaches, of the type favoured by computers, are ineffective. Furthermore, the best Go players seem to rely upon intuition—a phenomenon we don't know how replicate in lines of programming code. Therefore, the argument went, humans could rest assured of their intellectual superiority over computers: we might no longer be better chess players but we would always be better Go players. Except, of course, that's not how it turned out.

In October 2015, AlphaGo—a program developed by the AI company Google DeepMind—beat a human professional Go player. Five months later, the program beat a human player of the top rank in a five-game match. And in May 2017, AlphaGo beat the world's number one human player in a three-game match. This *was* an achievement worth noting. Whereas Deep had followed a script for chess and was successful simply because its raw processing power enabled it to follow the script quickly, AlphaGo *learned* to be successful at Go. It did so by studying a database of 30 million moves made by human experts in games—essentially, the program was made to watch thousands of hours of recorded human play. The developers reinforced that training by making AlphaGo play against different instances of itself in order to generate and analyze new moves. This combination of deep learning and reinforcement learning was enough to create a Go program capable of crushing any human opponent.

The human factor was not entirely absent from AlphaGo: the program required a pre-existing database of human games from which to learn. But it turns out humans aren't needed even for that purpose. In October 2017, Google DeepMind released AlphaGo Zero. This iteration of the program didn't need to study past games. Instead, scientists programmed it with the rules of Go and then let it start; it played at random against itself and observed the results. After three days AlphaGo Zero played 4.9 million games. In that three-day period AlphaGo Zero accumulated more Go knowledge than humans had acquired since they began playing the game about 2500 years ago. Within three days AlphaGo Zero became the best Go player on the planet, and trounced the initial version of AlphaGo 100–0 in a hundred-game match. Checkers, chess, Go … computers crush all human competition. Using machine learning techniques, computers are capable of beating humans at *any* rule-based board game.

But so what? Does excellence in board games make computers intelligent? Computers can perform arithmetic better than people, too, but few of us would argue that makes them intelligent. To be clear: this is *not* to say developments such as AlphaGo Zero are unimportant; indeed, the techniques that went into the creation of AlphaGo Zero might soon change the nature of our civilisation. But they aren't necessarily a signifier of human-level intelligence. We don't revere the names of Einstein, Shakespeare, and Van Gogh because they excelled at backgammon; the thought of judging our friends, family, and colleagues on the basis of an ability to strategise in checkers simply doesn't enter our heads. Surely computers must exhibit something more than a prowess at chess if we are to consider them intelligent?

So why did those early AI pioneers think the development of a chess-playing computer, as in "Moxon's Master", would light the path to artificial intelligence? Perhaps the answer lies in part because those AI pioneers were themselves exceptionally intelligent. Often such people spend a large part of their life lost in their thoughts. They reason, pontificate, imagine, deduce, argue, contemplate ... and the body becomes merely a means of getting the brain from A to B. But of course being fully human isn't solely about intellect. How could it be? We are biological creatures, and hundreds of millions of years of evolution ensures that the human body influences the thoughts and feelings passing through the brain to which it's attached. It makes no sense to talk about human-level intelligence as if it were entirely separate from embodied experience.

The commentators who argued the automaton in "Moxon's Master" must be intelligent because it played a decent game of chess were, I believe, misguided. The automaton possessed no more intelligence than Deep Blue. However, perhaps one could argue the automaton was intelligent, or at least possessed a mind capable of self-awareness, on the basis it wasn't a very good chess player—*and was angry about the fact!* The automaton killed Moxon in a fit of pique, a rage caused by a failure to foresee Moxon's checkmating move. Anger might not be attractive, and it often leads to behaviour we would classify as stupid, but it's an element of human intelligence. Similarly, all those other emotions that move us, whether positively or negatively, are an inevitable part of the mix. When attempting to define what we mean by mind surely we shouldn't pretend emotions are irrelevant and then extract one intellectual pursuit—chess, arithmetic, or whatever—and conclude that excellence in this single domain is a token of general intelligence or self-awareness?

From what source, however, would Moxon's automaton or any other machine experience emotion? Forget chess—put the automaton in a room with one-way mirrors and play it some catchy music. If it dances as if no one is looking then I'll grant it might be intelligent. But in the absence of glands, hormones, drives that originate in our mammalian and even pre-mammalian ancestors, a machine would simply stand still and do what it was programmed to do. Wouldn't it?

My dance test is impractical, so let's return to more mainstream thinking.

*   *   *

What, then, *is* artificial intelligence, if it isn't the ability for a machine to beat a person at chess or Go?

One hugely influential contribution to the AI debate came from Alan Turing. The famous codebreaker was interested in the question: can machines think? It's hard to get a handle on this question because of the difficulty, as we've just discussed, of defining precisely what we mean by "think". (Was Deep Blue "thinking" when it beat Kasparov?) However, around the same time he was writing his chess program, Turing came up with a closely related question that *can* be answered by experiment. The so-called "Turing test" has different versions and interpretations, but the general flavour is as follows.

Suppose a person converses naturally with two test participants—one human, one machine, both of whom are hidden from view—and at the end of the interrogation is unable to distinguish the machine from the human. In other words, suppose the machine can imitate a human so well its responses are indistinguishable from those that might be given by a human. Under those circumstances the interrogator should grant that the machine can "think".

But when you think about it—and I'm granting that you, the reader, can think—the Turing test has some obvious shortcomings. At least, it does if you are interested in measuring a machine's intelligence.

For example, people often exhibit unintelligent behaviour (don't get me started!) so an intelligent machine that couldn't mimic *unintelligent* behaviour would eventually fail the test. Similarly, a super-intelligent machine would be able to do things humans can't; unless it "dumbed down", and refused to show those behaviours, it would fail the test.

Another shortcoming of this approach: human interrogators might assign qualities to a machine because of our inherent tendency to anthropomorphise. An example of this from my own experience occurred back in the 1980s. I set my cheap chess program to run at low-intermediate level; sometimes I beat it, sometimes it beat me. On its turn to move the program would sometimes pause—and in those cases I couldn't help but believe I was playing against a conscious entity that was "thinking" about the best move to make. I now know those pauses were inserted deliberately by the code, so human opponents wouldn't lose heart. But back then I was a victim to anthropomorphism. Garry Kasparov, when he played Deep Blue in 1997, made a similar mistake (albeit one made when he was operating at a completely different level to me). On move 44 of game 1 the program sacrificed a piece to try and gain a short-term advantage, the sort of illogical move a human player might make. Kasparov won the game, but move 44 spooked him: he thought he saw in it a sign of "superior intelligence". The reality was more mundane. The algorithm was unable to calculate the best move in this position, and so a programmed safeguard kicked in—Deep Blue produced a move at random. Kasparov anthropomorphised by seeing a programming bug as a sign of intelligence. This tendency to anthropomorphise isn't restricted to chess. Religion, for example, provides yet another case study: throughout history worshippers of various faiths have bestowed personhood (or even some form of divinity) to non-humans. Anthropomorphism occurs in all walks of life.

In any case, the Turing test doesn't address the deep problem: a computer might be able to pass the test by mechanically following a script. Indeed, some readers of this book might have interacted with a chatbot—a program capable of conversation, usually via text—without realizing it. Chatbots are used increasingly on customer service websites, and if the query that drove you to the website is easily handled then you probably completed the conversation by thanking the program and failing entirely to notice that you were interacting with code. But even if a chatbot is defined to be "thinking" because it passes the Turing test it doesn't follow that it's conscious or acting with intent.

Perhaps the sensible approach to the question of AI, at least at our present stage, is not to get too hung up on matters of philosophy. Perhaps we should just write programs, build machines, and simply watch how they get on. The approach exemplified by AlphaGo Zero, for example, works brilliantly for clearly defined worlds possessing simple rules that can be pre-programmed. On the other hand, this approach is not necessarily going to work for complicated tasks in the real world; AlphaGo Zero would make a lousy taxi driver, for example. Complex, real-world tasks might require a quite different approach—but there's no reason to suppose AI practitioners won't develop those approaches. At the current rate of progress there'll come a time—a few years hence, perhaps a few decades—when AIs will be navigating the everyday world as effectively as humans. When that happens it will be difficult to argue those entities *aren't* intelligent. But their intelligence, I believe, will be quite alien. Their method of thinking will be entirely different to our own. And the future, perhaps, will lie in some type of marriage of our two different forms of intelligence.

* * *

This speculation naturally leads into the theme of Chapter 6: what will be the human response to human-like machines?

Notes and Further Reading

the famously mordant "Bitter" Bierce—Ambrose was a popular author, a feared literary critic, and an influential figure in American letters. His book *The Devil's Dictionary* (1911) is an acknowledged classic of American literature.

three different and diverging interpretations of "Moxon's Master"—For three different readings of the story, and in particular for speculations on who precisely Moxon's master was, see Bleiler (1985), Rottensteiner (1988), and Canty (1996).

chess is a game for the "brainy"—Chess history contains many examples of mental feats that seem barely possible. It's tempting to consider such feats as the being the province of a very few individuals in possession of almost superhuman brainpower. But studies have shown that chess grandmasters don't have a better working memory than the general population; rather, when it comes to chess, they see the board, and the relationships between pieces in active play, in a different way to the rest of us. A similar effect occurs in many different domains of expertise: experts employ more effective techniques than do non-experts, but it's mainly the result of practice rather than god-given talent. Conversely, ordinary members of the public can be trained to perform seemingly incredible feats of mental athletics. See, for example, Foer (2011).

The Turk, a chess-playing "automaton"—Standage (2002) describes the fascinating history of The Mechanical Turk.

Turing, by any reasonable definition of the term—Hodges (1983) is the classic biography of Turing. Since the publication of Hodges' book interest in Turing has grown, and numerous publications and websites have appeared. A good modern book about Turing's life and intellectual activities, aimed at a wide readership, is Copeland et al. (2017).

Deep Blue beat Garry Kasparov—Feng-hsiung Hsu, a Chinese computer scientist, was the principal designer of Deep Blue. For an account of how he built a machine that could beat the great Kasparov, see Hsu (2002).

The so-called "Turing test"—Warwick and Shah (2016) is an interesting review of Turing's imitation game.

some type of marriage of our two different forms of intelligence—Science fiction authors, of course, have long investigated artificial intelligence and what its development might mean for humanity. Asimov's robot stories (which include five novels as well as more than three dozen short stories, most of which have been anthologized in Asimov (1982)) are the best known, but numerous SF works—in all types of media—have considered the question. For a non-fiction discussion of how human and artificial intelligence might co-exist, see for example Markoff (2015).

Bibliography

Asimov, A.: The Complete Robot. Doubleday, New York (1982)

Bierce, A.: Moxon's master. The San Francisco Examiner. 16 April 1899

Bierce, A.: The Devil's Dictionary. Neale, New York (1911)

Bleiler, E.F.: Who was Moxon's master? Extrapolation. **26**(3), 181–189 (1985)

Canty, D.: The meaning of "Moxon's Master". Sci. Fict. Stud. **23**(3), 538–541 (1996)

Copeland, J., Bowen, J., Sprevak, M., Wilson, R.: The Turing Guide. Princeton University Press, Princeton, NJ (2017)

Foer, J.: Moonwalking with Einstein: The Art and Science of Remembering Everything. Penguin, New York (2011)

Hodges, A.: Alan Turing: The Enigma. Simon & Schuster, New York (1983)

Hsu, F.-h.: Behind Deep Blue: Building the Computer that Defeated the World Chess Champion. Princeton University Press, Princeton, NJ (2002)

Markoff, J.: Machines of Loving Grace: The Quest for Common Ground Between Humans and Robots. Ecco Press, New York (2015)

Rottensteiner, F.: Who was really Moxon's master? Sci. Fict. Stud. **15**(1), 107–112 (1988)

Standage, T.: The Mechanical Turk: The True Story of the Chess-Playing Machine That Fooled the World. Allen Lane, London (2002)

Warwick, K., Shah, H.: Turing's Imitation Game: Conversations with the Unknown. Cambridge University Press, Cambridge (2016)

6

Attractive Androids

The Sandman (E.T.A. Hoffmann)

NATHANAEL TO LOTHAIR

I know you are all very uneasy because I have not written for such a long, long time. Mother, to be sure, is angry, and Clara, I dare say, believes I am living here in riot and revelry, and quite forgetting my sweet angel, whose image is so deeply engraved upon my heart and mind. But that is not so; daily and hourly do I think of you all, and my lovely Clara's form comes to gladden me in my dreams, and smiles upon me with her bright eyes, as graciously as she used to do in the days when I went in and out amongst you. Oh! how could I write to you in the distracted state of mind in which I have been, and which, until now, has quite bewildered me! A terrible thing has happened to me. Dark forebodings of some awful fate threatening me are spreading themselves out over my head like black clouds, impenetrable to every friendly ray of sunlight. I must now tell you what has taken place; I must, that I see well enough, but only to think upon it makes the wild laughter burst from my lips. Oh! my dear, dear Lothair, what shall I say to make you feel, if only in an inadequate way, that that which happened to me a few days ago could thus really exercise such a hostile and disturbing influence upon my life? Oh that you were here to see for yourself! but now you will, I suppose, take me for a superstitious ghost-seer. In a word, the terrible thing which I have experienced, the fatal

© Springer Nature Switzerland AG 2019
S. Webb, *New Light Through Old Windows: Exploring Contemporary Science Through 12 Classic Science Fiction Tales*, Science and Fiction,
https://doi.org/10.1007/978-3-030-03195-4_6

effect of which I in vain exert every effort to shake off, is simply that some days ago, namely, on the 30th October, at twelve o'clock at noon, a dealer in weather-glasses came into my room and wanted to sell me one of his wares. I bought nothing, and threatened to kick him downstairs, whereupon he went away of his own accord.

You will conclude that it can only be very peculiar relations—relations intimately intertwined with my life—that can give significance to this event, and that it must be the person of this unfortunate hawker which has had such a very inimical effect upon me. And so it really is. I will summon up all my faculties in order to narrate to you calmly and patiently as much of the early days of my youth as will suffice to put matters before you in such a way that your keen sharp intellect may grasp everything clearly and distinctly, in bright and living pictures. Just as I am beginning, I hear you laugh and Clara say, "What's all this childish nonsense about!" Well, laugh at me, laugh heartily at me, pray do. But, good God! my hair is standing on end, and I seem to be entreating you to laugh at me in the same sort of frantic despair in which Franz Moor entreated Daniel to laugh him to scorn. But to my story.

Except at dinner we, *i.e.*, I and my brothers and sisters, saw but little of our father all day long. His business no doubt took up most of his time. After our evening meal, which, in accordance with an old custom, was served at seven o'clock, we all went, mother with us, into father's room, and took our places around a round table. My father smoked his pipe, drinking a large glass of beer to it. Often he told us many wonderful stories, and got so excited over them that his pipe always went out; I used then to light it for him with a spill, and this formed my chief amusement. Often, again, he would give us picture-books to look at, whilst he sat silent and motionless in his easy-chair, puffing out such dense clouds of smoke that we were all as it were enveloped in mist. On such evenings mother was very sad; and directly it struck nine she said, "Come, children! off to bed! Come! The 'sandman' is come I see." And I always did seem to hear something trampling upstairs with slow heavy steps; that must be the Sandman. Once in particular I was very much frightened at this dull trampling and knocking; as mother was leading us out of the room I asked her, "O mamma! but who is this nasty Sandman who always sends us away from papa? What does he look like?"

"There is no Sandman, my dear child," mother answered; "when I say the Sandman is come, I only mean that you are sleepy and can't keep your eyes open, as if somebody had put sand in them." This answer of mother's did not satisfy me; nay, in my childish mind the thought clearly unfolded itself that mother denied there was a Sandman only to prevent us being afraid,—why, I always heard him come upstairs. Full of curiosity to learn something more about this Sandman and what he had to do with us children, I at length asked the old woman who acted as my youngest sister's attendant, what sort of a man he was—the Sandman? "Why, 'thanael, darling, don't you know?" she replied. "Oh! he's a wicked man, who comes to little children when they won't go to bed and throws handfuls of sand in their eyes, so that they jump out of their heads all bloody; and he puts them into a bag and takes them to the half-moon as food for his little ones; and they sit there in the nest and have hooked beaks like owls, and they pick naughty little boys' and girls' eyes out with them." After this I formed in my own mind a horrible picture of the cruel Sandman. When anything came blundering upstairs at night I trembled with fear and dismay; and all that my mother could get out of me were the stammered words "The Sandman! the Sandman!" whilst the tears coursed down my cheeks. Then I ran into my bedroom, and the whole night through tormented myself with the terrible apparition of the Sandman. I was quite old enough to perceive that the old woman's tale about the Sandman and his little ones' nest in the half-moon couldn't be altogether true; nevertheless the Sandman continued to be for me a fearful incubus, and I was always seized with terror—my blood always ran cold, not only when I heard anybody come up the stairs, but when I heard anybody noisily open my father's room door and go in. Often he stayed away for a long season altogether; then he would come several times in close succession.

This went on for years, without my being able to accustom myself to this fearful apparition, without the image of the horrible Sandman growing any fainter in my imagination. His intercourse with my father began to occupy my fancy ever more and more; I was restrained from asking my father about him by an unconquerable shyness; but as the years went on the desire waxed stronger and stronger within me to fathom the mystery myself and to see the fabulous Sandman. He had been the means of disclosing to me the path of the wonderful and the adventurous, which so easily find lodgment in the mind of the child. I liked nothing better than to hear or read horrible stories of goblins, witches, Tom Thumbs, and so on; but always at the head of them all stood the Sandman, whose picture I scribbled in the most extraordinary and repulsive forms with both chalk and coal everywhere, on the tables, and cupboard doors, and walls. When I was ten years old my mother removed me from the nursery into a little chamber off the corridor not far from my father's room. We still had to withdraw hastily whenever, on the stroke of nine, the mysterious unknown was heard in the house. As I lay in my little chamber I could hear him go into father's room, and soon afterwards I fancied there was a fine and peculiar smelling steam spreading itself through the house. As my curiosity waxed stronger, my resolve to make somehow or other the Sandman's acquaintance took deeper root. Often when my mother had gone past, I slipped quickly out of my room into the corridor, but I could never see anything, for always before I could reach the place where I could get sight of him, the Sandman was well inside the door. At last, unable to resist the impulse any longer, I determined to conceal myself in father's room and there wait for the Sandman.

One evening I perceived from my father's silence and mother's sadness that the Sandman would come; accordingly, pleading that I was excessively tired, I left the room before nine o'clock and concealed myself in a hiding-place close beside the door. The street door creaked, and slow, heavy, echoing steps crossed the passage towards the stairs. Mother hurried past me with my brothers and sisters. Softly—softly—I opened father's room door. He sat as usual, silent and motionless, with his back towards it; he did not hear me; and in a moment I was in and behind a curtain drawn before my father's open wardrobe, which stood just inside the room. Nearer and nearer and nearer came the echoing footsteps. There was a strange coughing and shuffling and mumbling outside. My heart beat with expectation and fear. A quick step now close, close beside the door, a noisy rattle of the handle, and the door flies open with a bang. Recovering my courage with an effort, I take a cautious peep out. In the middle of the room in front of my father stands the Sandman, the bright light of the lamp falling full upon his face. The Sandman, the terrible Sandman, is the old advocate *Coppelius* who often comes to dine with us.

But the most hideous figure could not have awakened greater trepidation in my heart than this Coppelius did. Picture to yourself a large broad-shouldered man, with an immensely big head, a face the colour of yellow-ochre, grey bushy eyebrows, from beneath which two piercing, greenish, cat-like eyes glittered, and a prominent Roman nose hanging over his upper lip. His distorted mouth was often screwed up into a malicious smile; then two dark-red spots appeared on his cheeks, and a strange hissing noise proceeded from between his tightly clenched teeth. He always wore an ash-grey coat of an old-fashioned cut, a waistcoat of the same, and nether extremities to match, but black stockings and buckles set with stones on his shoes. His little wig scarcely extended beyond the crown of his head, his hair was curled round high up above his big red ears, and plastered to his temples with cosmetic, and a broad closed hair-bag stood out prominently from his neck, so that you could see the silver buckle that fastened his folded neck-cloth. Altogether he was a most disagreeable and horribly ugly figure; but what we children detested most of all was his big coarse hairy hands; we could never fancy anything that he had once touched. This he had noticed; and so, whenever our good mother quietly placed a piece of cake or sweet fruit on our plates, he delighted to touch it under some pretext or other, until the bright tears stood in our eyes, and from disgust and loathing we lost the enjoyment of the tit-bit that was intended to please us. And he did just the same thing when father gave us a glass of sweet wine on holidays. Then he would quickly pass his hand over it, or even sometimes raise the glass to his blue lips, and he laughed quite sardonically when all we dared do was to express our vexation in stifled sobs. He habitually called us the "little brutes;" and when he was present we might not utter a sound; and we cursed the ugly spiteful man who deliberately and intentionally spoilt all our little pleasures. Mother seemed to dislike this hateful Coppelius as much as we did; for as soon as he appeared her cheerfulness and bright and natural manner were transformed into sad, gloomy seriousness. Father treated him as if he were a being of some higher race, whose ill-manners were to be tolerated, whilst no efforts ought to be spared to keep him in good-humour. He had only to give a slight hint, and his favourite dishes were cooked for him and rare wine uncorked.

As soon as I saw this Coppelius, therefore, the fearful and hideous thought arose in my mind that he, and he alone, must be the Sandman; but I no longer conceived of the Sandman as the bugbear in the old nurse's fable, who fetched children's eyes and took them to the half-moon as food for his little ones—no! but as an ugly spectre-like fiend bringing trouble and misery and ruin, both temporal and everlasting, everywhere wherever he appeared.

I was spell-bound on the spot. At the risk of being discovered, and, as I well enough knew, of being severely punished, I remained as I was, with my head thrust through the curtains listening. My father received Coppelius in a ceremonious manner. "Come, to work!" cried the latter, in a hoarse snarling voice, throwing off his coat. Gloomily and silently my father took off his dressing-gown, and both put on long black smock-frocks. Where they took them from I forgot to notice. Father opened the folding-doors of a cupboard in the wall; but I saw that what I had so long taken to be a cupboard was really a dark recess, in which was a little hearth. Coppelius approached it, and a blue flame crackled upwards from it. Round about were all kinds of strange utensils. Good God! as my old father bent down over the fire how different he looked! His gentle and venerable features seemed to be drawn up by some dreadful convulsive pain into an ugly, repulsive Satanic mask. He looked like Coppelius. Coppelius plied the red-hot tongs and drew bright glowing masses out of the thick smoke and began assiduously to hammer them. I fancied that there were men's faces visible round about, but without eyes, having ghastly deep black holes where the eyes should have been. "Eyes here! Eyes here!" cried Coppelius, in a hollow sepulchral voice. My blood ran cold with horror; I screamed and tumbled out of my hiding-place into the floor. Coppelius immediately seized upon me. "You little brute! You little brute!" he bleated, grinding his teeth. Then, snatching me up, he threw me on the hearth, so that the flames began to singe my hair. "Now we've got eyes—eyes—a beautiful pair of children's eyes," he whispered, and, thrusting his hands into the flames he took out some red-hot grains and was about to strew them into my eyes. Then my father clasped his hands and entreated him, saying, "Master, master, let my Nathanael keep his eyes—oh! do let him keep them." Coppelius laughed shrilly and replied, "Well then, the boy may keep his eyes and whine and pule his way through the world; but we will now at any rate observe the mechanism of the hand and the foot." And therewith he roughly laid hold upon me, so that my joints cracked, and twisted my hands and my feet, pulling them now this way, and now that, "That's not quite right altogether! It's better as it was!—the old fellow knew what he was about." Thus lisped and hissed Coppelius; but all around me grew black and dark; a sudden convulsive pain shot through all my nerves and bones; I knew nothing more.

I felt a soft warm breath fanning my cheek; I awakened as if out of the sleep of death; my mother was bending over me. "Is the Sandman still there?" I stammered. "No, my dear child; he's been gone a long, long time; he'll not hurt you." Thus spoke my mother, as she kissed her recovered darling and pressed him to her heart. But why should I tire you, my dear Lothair? why do I dwell at such length on these details, when there's so much remains to be said? Enough—I was detected in my eavesdropping, and roughly handled by Coppelius. Fear and terror had brought on a violent fever, of which I lay ill several weeks. "Is the Sandman still there?" these were the first words I uttered on coming to myself again, the first sign of my recovery, of my safety. Thus, you see, I have only to relate to you the most terrible moment of my youth for you to thoroughly understand that it must not be ascribed to the weakness of my eyesight if all that I see is colourless, but to the fact that a mysterious destiny has hung a dark veil of clouds about my life, which I shall perhaps only break through when I die.

Coppelius did not show himself again; it was reported he had left the town. It was about a year later when, in pursuance of the old unchanged custom, we sat around the round table in the evening. Father was in very good spirits, and was telling us amusing tales about his youthful travels. As it was striking nine we all at once heard the street door creak on its hinges, and slow ponderous steps echoed across the passage and up the stairs. "That is Coppelius," said my mother, turning pale. "Yes, it is Coppelius," replied my father in a faint broken voice. The tears started from my mother's eyes. "But, father, father," she cried, "must it be so?"

"This is the last time," he replied; "this is the last time he will come to me, I promise you. Go now, go and take the children. Go, go to bed—good-night."

As for me, I felt as if I were converted into cold, heavy stone; I could not get my breath. As I stood there immovable my mother seized me by the arm. "Come, Nathanael! do come along!" I suffered myself to be led away; I went into my room. "Be a good boy and keep quiet," mother called after me; "get into bed and go to sleep." But, tortured by indescribable fear and uneasiness, I could not close my eyes. That hateful, hideous Coppelius stood before me with his glittering eyes, smiling maliciously down upon me; in vain did I strive to banish the image. Somewhere about midnight there was a terrific crack, as if a cannon were being fired off. The whole house shook; something went rustling and clattering past my door; the house-door was pulled to with a bang. "That is Coppelius," I cried, terror-struck, and leapt out of bed. Then I heard a wild heartrending scream; I rushed into my father's room; the door stood open, and clouds of suffocating smoke came rolling towards me. The servant-maid shouted, "Oh! my master! my master!" On the floor in front of the smoking hearth lay my father, dead, his face burned black and fearfully distorted, my sisters weeping and moaning around him, and my mother lying near them in a swoon. "Coppelius, you atrocious fiend, you've killed my father," I shouted. My senses left me. Two days later, when my father was placed in his coffin, his features were mild and gentle again as they had been when he was alive. I found great consolation in the thought that his association with the diabolical Coppelius could not have ended in his everlasting ruin.

Our neighbours had been awakened by the explosion; the affair got talked about, and came before the magisterial authorities, who wished to cite Coppelius to clear himself. But he had disappeared from the place, leaving no traces behind him.

Now when I tell you, my dear friend, that the weather-glass hawker I spoke of was the villain Coppelius, you will not blame me for seeing impending mischief in his inauspicious reappearance. He was differently dressed; but Coppelius's figure and features are too deeply impressed upon my mind for me to be capable of making a mistake in the matter. Moreover, he has not even changed his name. He proclaims himself here, I learn, to be a Piedmontese mechanician, and styles himself Giuseppe Coppola.

I am resolved to enter the lists against him and revenge my father's death, let the consequences be what they may.

Don't say a word to mother about the reappearance of this odious monster. Give my love to my darling Clara; I will write to her when I am in a somewhat calmer frame of mind.

Adieu, &c.

*   *   *

CLARA TO NATHANAEL

You are right, you have not written to me for a very long time, but neverthe-less I believe that I still retain a place in your mind and thoughts. It is a proof that you were thinking a good deal about me when you were sending off your last letter to brother Lothair, for instead of directing it to him you directed it to me. With joy I tore open the envelope, and did not perceive the mistake until I read the words, "Oh! my dear, dear Lothair." Now I know I ought not to have read any more of the letter, but ought to have given it to my brother. But as you have so often in innocent raillery made it a sort of reproach against me that I possessed such a calm, and, for a woman, cool-headed temperament that I should be like the woman we read of—if the house was threatening to tumble down, I should, before hastily fleeing, stop to smooth down a crumple in the window-curtains—I need hardly tell you that the beginning of your letter quite upset me. I could scarcely breathe; there was a bright mist before my eyes. Oh! my darling Nathanael! what could this terrible thing be that had happened? Separation from you—never to see you again, the thought was like a sharp knife in my heart. I read on and on. Your description of that horrid Coppelius made my flesh creep. I now learnt for the first time what a terrible and violent death your good old father died. Brother Lothair, to whom I handed over his property, sought to comfort me, but with little success. That horrid weather-glass hawker Giuseppe Coppola followed me everywhere; and I am almost ashamed to confess it, but he was able to disturb my sound and in general calm sleep with all sorts of wonderful dream-shapes. But soon—the next day—I saw everything in a different light. Oh! do not be angry with me, my best-beloved, if, despite your strange presentiment that Coppelius will do you some mischief, Lothair tells you I am in quite as good spirits, and just the same as ever.

I will frankly confess, it seems to me that all that was fearsome and terrible of which you speak, existed only in your own self, and that the real true outer world had but little to do with it. I can quite admit that old Coppelius may have been highly obnoxious to you children, but your real detestation of him arose from the fact that he hated children.

Naturally enough the gruesome Sandman of the old nurse's story was asso-ciated in your childish mind with old Coppelius, who, even though you had not believed in the Sandman, would have been to you a ghostly bugbear, especially dangerous to children. His mysterious labours along with your father at night-time were, I daresay, nothing more than secret experiments in alchemy, with which your mother could not be over well pleased, owing to the large sums of money that most likely were thrown away upon them; and besides, your father, his mind full of the deceptive striving after higher knowledge, may probably have become rather indifferent to his family, as so often happens in the case of such experimentalists. So also it is equally probable that your father brought about his death by his own imprudence, and that Coppelius is not to blame for it. I must tell you that yesterday I asked our experienced neighbour, the chemist, whether in experiments of this kind an explosion could take place which would have a momentarily fatal effect. He said, "Oh, certainly!" and described to me in his prolix and circumstantial way how it could be occasioned, mentioning at the same time so many strange and funny words that I could not remember them at all. Now I know you will be angry at your Clara, and will say, "Of the Mysterious which often clasps man in its invisible arms there's not a ray can find its way into this cold heart. She sees only the varied surface of the things of the world, and, like the little child, is pleased with the golden glittering fruit; at the kernel of which lies the fatal poison."

Oh! my beloved Nathanael, do you believe then that the intuitive prescience of a dark power working within us to our own ruin cannot exist also in minds which are cheerful, natural, free from care? But please forgive me that I, a simple girl, presume in any way to indicate to you what I really think of such an inward strife. After all, I should not find the proper words, and you would only laugh at me, not because my thoughts were stupid, but because I was so foolish as to attempt to tell them to you.

If there is a dark and hostile power which traitorously fixes a thread in our hearts in order that, laying hold of it and drawing us by means of it along a dangerous road to ruin, which otherwise we should not have trod—if, I say, there is such a power, it must assume within us a form like ourselves, nay, it must be ourselves; for only in that way can we believe in it, and only so understood do we yield to it so far that it is able to accomplish its secret purpose. So long as we have sufficient firmness, fortified by cheerfulness, to always acknowledge foreign hostile influences for what they really are, whilst we quietly pursue the path pointed out to us by both inclination and calling, then this mysterious power perishes in its futile struggles to attain the form which is to be the reflected image of ourselves. It is also certain, Lothair adds, that if we have once voluntarily given ourselves up to this dark physical power, it often reproduces within us the strange forms which the outer world throws in our way, so that thus it is we ourselves who engender within ourselves the spirit which by some remarkable delusion we imagine to speak in that outer form. It is the phantom of our own self whose intimate relationship with, and whose powerful influence upon our soul either plunges us into hell or elevates us to heaven. Thus you will see, my beloved Nathanael, that I and brother Lothair have well talked over the subject of dark powers and forces; and now, after I have with some difficulty written down the principal results of our discussion, they seem to me to contain many really profound thoughts. Lothair's last words, however, I don't quite understand altogether; I only dimly guess what he means; and yet I cannot help thinking it is all very true, I beg you, dear, strive to forget the ugly advocate Coppelius as well as the weather-glass hawker Giuseppe Coppola. Try and convince yourself that these foreign influences can have no power over you, that it is only the belief in their hostile power which can in reality make them dangerous to you. If every line of your letter did not betray the violent excitement of your mind, and if I did not sympathise with your condition from the bottom of my heart, I could in truth jest about the advocate Sandman and weather-glass hawker Coppelius. Pluck up your spirits! Be cheerful! I have resolved to appear to you as your guardian-angel if that ugly man Coppola should dare take it into his head to bother you in your dreams, and drive him away with a good hearty laugh. I'm not afraid of him and his nasty hands, not the least little bit; I won't let him either as advocate spoil any dainty tit-bit I've taken, or as Sandman rob me of my eyes.

My darling, darling Nathanael,

Eternally your, &c. &c.

* * *

NATHANAEL TO LOTHAIR

I am very sorry that Clara opened and read my last letter to you; of course the mistake is to be attributed to my own absence of mind. She has written me a very deep philosophical letter, proving conclusively that Coppelius and Coppola only exist in my own mind and are phantoms of my own self, which will at once be dissipated, as soon as I look upon them in that light. In very truth one can hardly believe that the mind which so often sparkles in those bright, beautifully smiling, childlike eyes of hers like a sweet lovely dream could draw such subtle and scholastic distinctions. She also mentions your name. You have been talking about me. I suppose you have been giving her lectures, since she sifts and refines everything so acutely. But enough of this! I must now tell you it is most certain that the weather-glass hawker Giuseppe Coppola is not the advocate Coppelius. I am attending the lectures of our recently appointed Professor of Physics, who, like the distinguished naturalist, is called Spalanzani, and is of Italian origin. He has known Coppola for many years; and it is also easy to tell from his accent that he really is a Piedmontese. Coppelius was a German, though no honest German, I fancy. Nevertheless I am not quite satisfied. You and Clara will perhaps take me for a gloomy dreamer, but nohow can I get rid of the impression which Coppelius's cursed face made upon me. I am glad to learn from Spalanzani that he has left the town. This Professor Spalanzani is a very queer fish. He is a little fat man, with prominent cheek-bones, thin nose, projecting lips, and small piercing eyes. You cannot get a better picture of him than by turning over one of the Berlin pocket-almanacs and looking at Cagliostro's portrait engraved by Chodowiecki; Spalanzani looks just like him.

Once lately, as I went up the steps to his house, I perceived that beside the curtain which generally covered a glass door there was a small chink. What it was that excited my curiosity I cannot explain; but I looked through. In the room I saw a female, tall, very slender, but of perfect proportions, and splendidly dressed, sitting at a little table, on which she had placed both her arms, her hands being folded together. She sat opposite the door, so that I could easily see her angelically beautiful face. She did not appear to notice me, and there was moreover a strangely fixed look about her eyes, I might almost say they appeared as if they had no power of vision; I thought she was sleeping with her eyes open. I felt quite uncomfortable, and so I slipped away quietly into the Professor's lecture-room, which was close at hand. Afterwards I learnt that the figure which I had seen was Spalanzani's daughter, Olimpia, whom he keeps locked in a most wicked and unaccountable way, and no man is ever allowed to come near her. Perhaps, however, there is after all, something peculiar about her; perhaps she's an idiot or something of that sort. But why am I telling you all this? I could have told you it all better and more in detail when I see you. For in a fortnight I shall be amongst you. I must see my dear sweet angel, my Clara, again. Then the little bit of ill-temper, which, I must confess, took possession of me after her fearfully sensible letter, will be blown away. And that is the reason why I am not writing to her as well to-day.

With all best wishes, &c.

* * *

Nothing more strange and extraordinary can be imagined, gracious reader, than what happened to my poor friend, the young student Nathanael, and which I have undertaken to relate to you. Have you ever lived to experience anything that completely took possession of your heart and mind and thoughts to the utter exclusion of everything else? All was seething and boiling within you; your blood, heated to fever pitch, leapt through your veins and inflamed your cheeks. Your gaze was so peculiar, as if seeking to grasp in empty space forms not seen of any other eye, and all your words ended in sighs betokening some mystery. Then your friends asked you, "What is the matter with you, my dear friend? What do you see?" And, wishing to describe the inner pictures in all their vivid colours, with their lights and their shades, you in vain struggled to find words with which to express yourself. But you felt as if you must gather up all the events that had happened, wonderful, splendid, terrible, jocose, and awful, in the very first word, so that the whole might be revealed by a single electric discharge, so to speak. Yet every word and all that partook of the nature of communication by intelligible sounds seemed to be colourless, cold, and dead. Then you try and try again, and stutter and stammer, whilst your friends' prosy questions strike like icy winds upon your heart's hot fire until they extinguish it. But if, like a bold painter, you had first sketched in a few audacious strokes the outline of the picture you had in your soul, you would then easily have been able to deepen and intensify the colours one after the other, until the varied throng of living figures carried your friends away, and they, like you, saw themselves in the midst of the scene that had proceeded out of your own soul.

Strictly speaking, indulgent reader, I must indeed confess to you, nobody has asked me for the history of young Nathanael; but you are very well aware that I belong to that remarkable class of authors who, when they are bearing anything about in their minds in the manner I have just described, feel as if everybody who comes near them, and also the whole world to boot, were asking, "Oh! what is it? Oh! do tell us, my good sir?" Hence I was most powerfully impelled to narrate to you Nathanael's ominous life. My soul was full of the elements of wonder and extraordinary peculiarity in it; but, for this very reason, and because it was necessary in the very beginning to dispose you, indulgent reader, to bear with what is fantastic—and that is not a little thing— I racked my brain to find a way of commencing the story in a significant and original manner, calculated to arrest your attention. To begin with "Once upon a time," the best beginning for a story, seemed to me too tame; with "In the small country town S—— lived," rather better, at any rate allowing plenty of room to work up to the climax; or to plunge at once in medias res, "'Go to the devil!' cried the student Nathanael, his eyes blazing wildly with rage and fear, when the weather-glass hawker Giuseppe Coppola"—well, that is what I really had written, when I thought I detected something of the ridiculous in Nathanael's wild glance; and the history is anything but laughable. I could not find any words which seemed fitted to reflect in even the feeblest degree the brightness of the colours of my mental vision. I determined not to begin at all. So I pray you, gracious reader, accept the three letters which my friend Lothair has been so kind as to communicate to me as the outline of the picture, into which I will endeavour to introduce more and more colour as I proceed with my narrative. Perhaps, like a good portrait-painter, I may succeed in depicting more than one figure in such wise that you will recognise it as a good likeness without being acquainted with the original, and feel as if you had very often seen the original with your own bodily eyes. Perhaps, too, you will then believe that nothing is more wonderful, nothing more fantastic than real life, and that all that a writer can do is to present it as a dark reflection from a dim cut mirror.

In order to make the very commencement more intelligible, it is necessary to add to the letters that, soon after the death of Nathanael's father, Clara and Lothair, the children of a distant relative, who had likewise died, leaving them orphans, were taken by Nathanael's mother into her own house. Clara and Nathanael conceived a warm affection for each other, against which not the slightest objection in the world could be urged. When therefore Nathanael left home to prosecute his studies in G——, they were betrothed. It is from G—— that his last letter is written, where he is attending the lectures of Spalanzani, the distinguished Professor of Physics.

I might now proceed comfortably with my narration, did not at this moment Clara's image rise up so vividly before my eyes that I cannot turn them away from it, just as I never could when she looked upon me and smiled so sweetly. Nowhere would she have passed for beautiful; that was the unanimous opinion of all who professed to have any technical knowledge of beauty. But whilst architects praised the pure proportions of her figure and form, painters averred that her neck, shoulders, and bosom were almost too chastely modelled, and yet, on the other hand, one and all were in love with her glorious Magdalene hair, and talked a good deal of nonsense about Battoni-like colouring. One of them, a veritable romanticist, strangely enough likened her eyes to a lake by Ruisdael, in which is reflected the pure azure of the cloudless sky, the beauty of woods and flowers, and all the bright and varied life of a living landscape. Poets and musicians went still further and said, "What's all this talk about seas and reflections? How can we look upon the girl without feeling that wonderful heavenly songs and melodies beam upon us from her eyes, penetrating deep down into our hearts, till all becomes awake and throbbing with emotion? And if we cannot sing anything at all passable then, why, we are not worth much; and this we can also plainly read in the rare smile which flits around her lips when we have the hardihood to squeak out something in her presence which we pretend to call singing, in spite of the fact that it is nothing more than a few single notes confusedly linked together." And it really was so. Clara had the powerful fancy of a bright, innocent, unaffected child, a woman's deep and sympathetic heart, and an understanding clear, sharp, and discriminating. Dreamers and visionaries had but a bad time of it with her; for without saying very much—she was not by nature of a talkative disposition—she plainly asked, by her calm steady look, and rare ironical smile, "How can you imagine, my dear friends, that I can take these fleeting shadowy images for true living and breathing forms?" For this reason many found fault with her as being cold, prosaic, and devoid of feeling; others, however, who had reached a clearer and deeper conception of life, were extremely fond of the intelligent, childlike, large-hearted girl But none had such an affection for her as Nathanael, who was a zealous and cheerful cultivator of the fields of science and art. Clara clung to her lover with all her heart; the first clouds she encountered in life were when he had to separate from her. With what delight did she fly into his arms when, as he had promised in his last letter to Lothair, he really came back to his native town and entered his mother's room! And as Nathanael had foreseen, the moment he saw Clara again he no longer thought about either the advocate Coppelius or her sensible letter; his ill-humour had quite disappeared.

Nevertheless Nathanael was right when he told his friend Lothair that the repulsive vendor of weather-glasses, Coppola, had exercised a fatal and disturbing influence upon his life. It was quite patent to all; for even during the first few days he showed that he was completely and entirely changed. He gave himself up to gloomy reveries, and moreover acted so strangely; they had never observed anything at all like it in him before. Everything, even his own life, was to him but dreams and presentiments. His constant theme was that every man who delusively imagined himself to be free was merely the plaything of the cruel sport of mysterious powers, and it was vain for man to resist them; he must humbly submit to whatever destiny had decreed for him. He went so far as to maintain that it was foolish to believe that a man could do anything in art or science of his own accord; for the inspiration in which alone any true artistic work could be done did not proceed from the spirit within outwards, but was the result of the operation directed inwards of some Higher Principle existing without and beyond ourselves.

This mystic extravagance was in the highest degree repugnant to Clara's clear intelligent mind, but it seemed vain to enter upon any attempt at refutation. Yet when Nathanael went on to prove that Coppelius was the Evil Principle which had entered into him and taken possession of him at the time he was listening behind the curtain, and that this hateful demon would in some terrible way ruin their happiness, then Clara grew grave and said, "Yes,

Nathanael. You are right; Coppelius is an Evil Principle; he can do dreadful things, as bad as could a Satanic power which should assume a living physical form, but only—only if you do not banish him from your mind and thoughts. So long as you believe in him he exists and is at work; your belief in him is his only power." Whereupon Nathanael, quite angry because Clara would only grant the existence of the demon in his own mind, began to dilate at large upon the whole mystic doctrine of devils and awful powers, but Clara abruptly broke off the theme by making, to Nathanael's very great disgust, some quite commonplace remark. Such deep mysteries are sealed books to cold, unsusceptible characters, he thought, without being clearly conscious to himself that he counted Clara amongst these inferior natures, and accordingly he did not remit his efforts to initiate her into these mysteries. In the morning, when she was helping to prepare breakfast, he would take his stand beside her, and read all sorts of mystic books to her, until she begged him—"But, my dear Nathanael, I shall have to scold you as the Evil Principle which exercises a fatal influence upon my coffee. For if I do as you wish, and let things go their own way, and look into your eyes whilst you read, the coffee will all boil over into the fire, and you will none of you get any breakfast." Then Nathanael hastily banged the book to and ran away in great displeasure to his own room.

Formerly he had possessed a peculiar talent for writing pleasing, sparkling tales, which Clara took the greatest delight in listening to; but now his productions were gloomy, unintelligible, and wanting in form, so that, although Clara out of forbearance towards him did not say so, he nevertheless felt how very little interest she took in them. There was nothing that Clara disliked so much as what was tedious; at such times her intellectual sleepiness was not to be overcome; it was betrayed both in her glances and in her words. Nathanael's effusions were, in truth, exceedingly tedious. His ill-humour at Clara's cold prosaic temperament continued to increase; Clara could not conceal her distaste of his dark, gloomy, wearying mysticism; and thus both began to be more and more estranged from each other without exactly being aware of it themselves. The image of the ugly Coppelius had, as Nathanael was obliged to confess to himself, faded considerably in his fancy, and it often cost him great pains to present him in vivid colours in his literary efforts, in which he played the part of the ghoul of Destiny. At length it entered into his head to make his dismal presentiment that Coppelius would ruin his happiness the subject of a poem. He made himself and Clara, united by true love, the central figures, but represented a black hand as being from time to time thrust into their life and plucking out a joy that had blossomed for them. At length, as they were standing at the altar, the terrible Coppelius appeared and touched Clara's lovely eyes, which leapt into Nathanael's own bosom, burning and hissing like bloody sparks. Then Coppelius laid hold upon him, and hurled him into a blazing circle of fire, which spun round with the speed of a whirlwind, and, storming and blustering, dashed away with him. The fearful noise it made was like a furious hurricane lashing the foaming sea-waves until they rise up like black, white-headed giants in the midst of the raging struggle. But through the midst of the savage fury of the tempest he heard Clara's voice calling, "Can you not see me, dear? Coppelius has deceived you; they were not my eyes which burned so in your bosom; they were fiery drops of your own heart's blood. Look at me, I have got my own eyes still." Nathanael thought, "Yes, that is Clara, and I am hers for ever." Then this thought laid a powerful grasp upon the fiery circle so that it stood still, and the riotous turmoil died away rumbling down a dark abyss. Nathanael looked into Clara's eyes; but it was death whose gaze rested so kindly upon him.

Whilst Nathanael was writing this work he was very quiet and sober-minded; he filed and polished every line, and as he had chosen to submit himself to the limitations of metre, he did not rest until all was pure and musical. When, however, he had at length finished it and read it aloud to himself he was seized with horror and awful dread, and he screamed, "Whose hideous voice is this?" But he soon came to see in it again nothing beyond a very successful poem, and he confidently believed it would enkindle Clara's cold temperament, though to what end she should be thus aroused was not quite clear to his own mind, nor yet what would be the real purpose served by tormenting her with these dreadful pictures, which prophesied a terrible and ruinous end to her affection.

Nathanael and Clara sat in his mother's little garden. Clara was bright and cheerful, since for three entire days her lover, who had been busy writing his poem, had not teased her with his dreams or forebodings. Nathanael, too, spoke in a gay and vivacious way of things of merry import, as he formerly used to do, so that Clara said, "Ah! now I have you again. We have driven away that ugly Coppelius, you see." Then it suddenly occurred to him that he had got the poem in his pocket which he wished to read to her. He at once took out the manuscript and began to read. Clara, anticipating something tedious as usual, prepared to submit to the infliction, and calmly resumed her knitting. But as the sombre clouds rose up darker and darker she let her knitting fall on her lap and sat with her eyes fixed in a set stare upon Nathanael's face. He was quite carried away by his own work, the fire of enthusiasm coloured his cheeks a deep red, and tears started from his eyes. At length he concluded, groaning and showing great lassitude; grasping Clara's hand, he sighed as if he were being utterly melted in inconsolable grief, "Oh! Clara! Clara!" She drew him softly to her heart and said in a low but very grave and impressive tone, "Nathanael, my darling Nathanael, throw that foolish, sense-less, stupid thing into the fire." Then Nathanael leapt indignantly to his feet, crying, as he pushed Clara from him, "You damned lifeless automaton!" and rushed away. Clara was cut to the heart, and wept bitterly. "Oh! he has never loved me, for he does not understand me," she sobbed.

Lothair entered the arbour. Clara was obliged to tell him all that had taken place. He was passionately fond of his sister; and every word of her complaint fell like a spark upon his heart, so that the displeasure which he had long entertained against his dreamy friend Nathanael was kindled into furious anger. He hastened to find Nathanael, and upbraided him in harsh words for his irrational behaviour towards his beloved sister. The fiery Nathanael answered him in the same style. "A fantastic, crack-brained fool," was retaliated with, "A miserable, common, everyday sort of fellow." A meeting was the inevitable consequence. They agreed to meet on the following morning behind the garden-wall, and fight, according to the custom of the students of the place, with sharp rapiers. They went about silent and gloomy; Clara had both heard and seen the violent quarrel, and also observed the fencing-master bring the rapiers in the dusk of the evening. She had a presentiment of what was to happen. They both appeared at the appointed place wrapped up in the same gloomy silence, and threw off their coats. Their eyes flaming with the bloodthirsty light of pugnacity, they were about to begin their contest when Clara burst through the garden door. Sobbing, she screamed, "You savage, terrible men! Cut me down before you attack each other; for how can I live when my lover has slain my brother, or my brother slain my lover?" Lothair let his weapon fall and gazed silently upon the ground, whilst Nathanael's heart was rent with sorrow, and all the affection which he had felt for his lovely Clara in the happiest days of her golden youth was awakened within him. His murderous weapon, too, fell from his hand; he threw himself at Clara's feet. "Oh! can you ever forgive me, my only, my dearly loved Clara? Can you, my dear brother Lothair, also forgive me?" Lothair was touched by his friend's great distress; the three young people embraced each other amidst endless tears, and swore never again to break their bond of love and fidelity.

Nathanael felt as if a heavy burden that had been weighing him down to the earth was now rolled from off him, nay, as if by offering resistance to the dark power which had possessed him, he had rescued his own self from the ruin which had threatened him. Three happy days he now spent amidst the loved ones, and then returned to G——, where he had still a year to stay before settling down in his native town for life.

Everything having reference to Coppelius had been concealed from the mother, for they knew she could not think of him without horror, since she as well as Nathanael believed him to be guilty of causing her husband's death.

* * *

When Nathanael came to the house where he lived he was greatly astonished to find it burnt down to the ground, so that nothing but the bare outer walls were left standing amidst a heap of ruins. Although the fire had broken out in the laboratory of the chemist who lived on the ground-floor, and had therefore spread upwards, some of Nathanael's bold, active friends had succeeded in time in forcing a way into his room in the upper storey and saving his books and manuscripts and instruments. They had carried them all uninjured into another house, where they engaged a room for him; this he now at once took possession of. That he lived opposite Professor Spalanzani did not strike him particularly, nor did it occur to him as anything more singular that he could, as he observed, by looking out of his window, see straight into the room where Olimpia often sat alone. Her figure he could plainly distinguish, although her features were uncertain and confused. It did at length occur to him, however, that she remained for hours together in the same position in which he had first discovered her through the glass door, sitting at a little table without any occupation whatever, and it was evident that she was constantly gazing across in his direction. He could not but confess to himself that he had never seen a finer figure. However, with Clara mistress of his heart, he remained perfectly unaffected by Olimpia's stiffness and apathy; and it was only occasionally that he sent a fugitive glance over his compendium across to her—that was all.

He was writing to Clara; a light tap came at the door. At his summons to "Come in," Coppola's repulsive face appeared peeping in. Nathanael felt his heart beat with trepidation; but, recollecting what Spalanzani had told him about his fellow-countryman Coppola, and what he had himself so faithfully promised his beloved in respect to the Sandman Coppelius, he was ashamed at himself for this childish fear of spectres. Accordingly, he controlled himself with an effort, and said, as quietly and as calmly as he possibly could, "I don't want to buy any weather-glasses, my good friend; you had better go elsewhere." Then Coppola came right into the room, and said in a hoarse voice, screwing up his wide mouth into a hideous smile, whilst his little eyes flashed keenly from beneath his long grey eyelashes, "What! Nee weather-gless? Nee weather-gless? 've got foine oyes as well—foine oyes!" Affrighted, Nathanael cried,

"You stupid man, how can you have eyes?—eyes—eyes?" But Coppola, laying aside his weather-glasses, thrust his hands into his big coat-pockets and brought out several spy-glasses and spectacles, and put them on the table. "Theer! Theer! Spect'cles! Spect'cles to put 'n nose! Them's my oyes—foine oyes." And he continued to produce more and more spectacles from his pockets until the table began to gleam and flash all over. Thousands of eyes were looking and blinking convulsively, and staring up at Nathanael; he could not avert his gaze from the table. Coppola went on heaping up his spectacles, whilst wilder and ever wilder burning flashes crossed through and through each other and darted their blood-red rays into Nathanael's breast. Quite overcome, and frantic with terror, he shouted, "Stop! stop! you terrible man!" and he seized Coppola by the arm, which he had again thrust into his pocket in order to bring out still more spectacles, although the whole table was covered all over with them. With a harsh disagreeable laugh Coppola gently freed himself; and with the words "So! went none! Well, here foine gless!" he swept all his spectacles together, and put them back into his coat-pockets, whilst from a breast-pocket he produced a great number of larger and smaller perspectives. As soon as the spectacles were gone Nathanael recovered his equanimity again; and, bending his thoughts upon Clara, he clearly discerned that the gruesome incubus had proceeded only from himself, as also that Coppola was a right honest mechanician and optician, and far from being Coppelius's dreaded double and ghost And then, besides, none of the glasses which Coppola now placed on the table had anything at all singular about them, at least nothing so weird as the spectacles; so, in order to square accounts with himself, Nathanael now really determined to buy something of the man. He took up a small, very beautifully cut pocket perspective, and by way of proving it looked through the window. Never before in his life had he had a glass in his hands that brought out things so clearly and sharply and distinctly. Involuntarily he directed the glass upon Spalanzani's room; Olimpia sat at the little table as usual, her arms laid upon it and her hands folded. Now he saw for the first time the regular and exquisite beauty of her features. The eyes, however, seemed to him to have a singular look of fixity and lifelessness. But as he continued to look closer and more carefully through the glass he fancied a light like humid moonbeams came into them. It seemed as if their power of vision was now being enkindled; their glances shone with ever-increasing vivacity. Nathanael remained standing at the window as if glued to the spot by a wizard's spell, his gaze rivetted unchangeably upon the divinely beautiful Olimpia. A coughing and shuffling of the feet awakened him out of his enchaining dream, as it were. Coppola stood behind him, "Tre zechini" (three ducats). Nathanael had completely forgotten the optician; he hastily paid the sum demanded. "Ain't 't? Foine gless? foine gless?" asked Coppola in his harsh unpleasant voice, smiling sardonically. "Yes, yes, yes," rejoined Nathanael impatiently; "adieu, my good friend." But Coppola did not leave the room without casting many peculiar side-glances upon Nathanael; and the young student heard him laughing loudly on the stairs. "Ah well!" thought he, "he's laughing at me because I've paid him too much for this little perspective— because I've given him too much money—that's it." As he softly murmured these words he fancied he detected a gasping sigh as of a dying man stealing awfully through the room; his heart stopped beating with fear. But to be sure he had heaved a deep sigh himself; it was quite plain. "Clara is quite right," said he to himself, "in holding me to be an incurable ghost-seer; and yet it's very ridiculous—ay, more than ridiculous, that the stupid thought of having paid Coppola too much for his glass should cause me this strange anxiety; I can't see any reason for it."

Now he sat down to finish his letter to Clara; but a glance through the window showed him Olimpia still in her former posture. Urged by an irresistible impulse he jumped up and seized Coppola's perspective; nor could he tear himself away from the fascinating Olimpia until his friend and brother Siegmund called for him to go to Professor Spalanzani's lecture. The curtains before the door of the all-important room were closely drawn, so that he could not see Olimpia. Nor could he even see her from his own room during the two following days, notwithstanding that he scarcely ever left his window, and maintained a scarce interrupted watch through Coppola's perspective upon her room. On the third day curtains even were drawn across the window. Plunged into the depths of despair,—goaded by longing and ardent desire, he hurried outside the walls of the town. Olimpia's image hovered about his path in the air and stepped forth out of the bushes, and peeped up at him with large and lustrous eyes from the bright surface of the brook. Clara's image was completely faded from his mind; he had no thoughts except for Olimpia. He uttered his love-plaints aloud and in a lachrymose tone, "Oh! my glorious, noble star of love, have you only risen to vanish again, and leave me in the darkness and hopelessness of night?"

Returning home, he became aware that there was a good deal of noisy bustle going on in Spalanzani's house. All the doors stood wide open; men were taking in all kinds of gear and furniture; the windows of the first floor were all lifted off their hinges; busy maid-servants with immense hair-brooms were driving backwards and forwards dusting and sweeping, whilst within could be heard the knocking and hammering of carpenters and upholsterers. Utterly astonished, Nathanael stood still in the street; then Siegmund joined him, laughing, and said, "Well, what do you say to our old Spalanzani?"

Nathanael assured him that he could not say anything, since he knew not what it all meant; to his great astonishment, he could hear, however, that they were turning the quiet gloomy house almost inside out with their dusting and cleaning and making of alterations. Then he learned from Siegmund that Spalanzani intended giving a great concert and ball on the following day, and that half the university was invited. It was generally reported that Spalanzani was going to let his daughter Olimpia, whom he had so long so jealously guarded from every eye, make her first appearance.

Nathanael received an invitation. At the appointed hour, when the carriages were rolling up and the lights were gleaming brightly in the decorated halls, he went across to the Professor's, his heart beating high with expectation. The company was both numerous and brilliant. Olimpia was richly and tastefully dressed. One could not but admire her figure and the regular beauty of her features. The striking inward curve of her back, as well as the wasp-like smallness of her waist, appeared to be the result of too-tight lacing. There was something stiff and measured in her gait and bearing that made an unfavourable impression upon many; it was ascribed to the constraint imposed upon her by the company. The concert began. Olimpia played on the piano with great skill; and sang as skilfully an *aria di bravura*, in a voice which was, if anything, almost too sharp, but clear as glass bells. Nathanael was transported with delight; he stood in the background farthest from her, and owing to the blinding lights could not quite distinguish her features. So, without being observed, he took Coppola's glass out of his pocket, and directed it upon the beautiful Olimpia. Oh! then he perceived how her yearning eyes sought him, how every note only reached its full purity in the loving glance which penetrated to and inflamed his heart. Her artificial *roulades* seemed to him to be the exultant cry towards heaven of the soul refined by love; and when at last, after the *cadenza*, the long trill rang shrilly and loudly through the hall, he felt as if he were suddenly grasped by burning arms and could no longer control himself,—he could not help shouting aloud in his mingled pain and delight, "Olimpia!" All eyes were turned upon him; many people laughed. The face of the cathedral organist wore a still more gloomy look than it had done before, but all he said was, "Very well!"

The concert came to an end, and the ball began. Oh! to dance with her—with her—that was now the aim of all Nathanael's wishes, of all his desires. But how should he have courage to request her, the queen of the ball, to grant him the honour of a dance? And yet he couldn't tell how it came about, just as the dance began, he found himself standing close beside her, nobody having as yet asked her to be his partner; so, with some difficulty stammering out a few words, he grasped her hand. It was cold as ice; he shook with an awful, frosty shiver. But, fixing his eyes upon her face, he saw that her glance was beaming upon him with love and longing, and at the same moment he thought that the pulse began to beat in her cold hand, and the warm life-blood to course through her veins. And passion burned more intensely in his own heart also; he threw his arm round her beautiful waist and whirled her round the hall. He had always thought that he kept good and accurate time in dancing, but from the perfectly rhythmical evenness with which Olimpia danced, and which frequently put him quite out, he perceived how very faulty his own time really was. Notwithstanding, he would not dance with any other lady; and everybody else who approached Olimpia to call upon her for a dance, he would have liked to kill on the spot. This, however, only happened twice; to his astonishment Olimpia remained after this without a partner, and he failed not on each occasion to take her out again. If Nathanael had been able to see anything else except the beautiful Olimpia, there would inevitably have been a good deal of unpleasant quarrelling and strife; for it was evident that Olimpia was the object of the smothered laughter only with difficulty suppressed, which was heard in various corners amongst the young people; and they followed her with very curious looks, but nobody knew for what reason. Nathanael, excited by dancing and the plentiful supply of wine he had consumed, had laid aside the shyness which at other times characterised him. He sat beside Olimpia, her hand in his own, and declared his love enthusiastically and passionately in words which neither of them understood, neither he nor Olimpia. And yet she perhaps did, for she sat with her eyes fixed unchangeably upon his, sighing repeatedly, "Ach! Ach! Ach!" Upon this Nathanael would answer, "Oh, you glorious heavenly lady! You ray from the promised paradise of love! Oh! what a profound soul you have! my whole being is mirrored in it!" and a good deal more in the same strain. But Olimpia only continued to sigh "Ach! Ach!" again and again.

Professor Spalanzani passed by the two happy lovers once or twice, and smiled with a look of peculiar satisfaction. All at once it seemed to Nathanael, albeit he was far away in a different world, as if it were growing perceptibly darker down below at Professor Spalanzani's. He looked about him, and to his very great alarm became aware that there were only two lights left burning in the hall, and they were on the point of going out. The music and dancing had long ago ceased. "We must part—part!" he cried, wildly and despairingly; he kissed Olimpia's hand; he bent down to her mouth, but ice-cold lips met his burning ones. As he touched her cold hand, he felt his heart thrilled with awe; the legend of "The Dead Bride" shot suddenly through his mind. But Olimpia had drawn him closer to her, and the kiss appeared to warm her lips into vitality. Professor Spalanzani strode slowly through the empty apartment, his footsteps giving a hollow echo; and his figure had, as the flickering shadows played about him, a ghostly, awful appearance. "Do you love me? Do you love me, Olimpia? Only one little word—Do you love me?" whispered Nathanael, but she only sighed, "Ach! Ach!" as she rose to her feet. "Yes, you are my lovely, glorious star of love," said Nathanael, "and will shine for ever, purifying and ennobling my heart"

"Ach! Ach!" replied Olimpia, as she moved along. Nathanael followed her; they stood before the Professor. "You have had an extraordinarily animated conversation with my daughter," said he, smiling; "well, well, my dear Mr. Nathanael, if you find pleasure in talking to the stupid girl, I am sure I shall be glad for you to come and do so." Nathanael took his leave, his heart singing and leaping in a perfect delirium of happiness.

During the next few days Spalanzani's ball was the general topic of conversation. Although the Professor had done everything to make the thing a splendid success, yet certain gay spirits related more than one thing that had occurred which was quite irregular and out of order. They were especially keen in pulling Olimpia to pieces for her taciturnity and rigid stiffness; in spite of her beautiful form they alleged that she was hopelessly stupid, and in this fact they discerned the reason why Spalanzani had so long kept her concealed from publicity. Nathanael heard all this with inward wrath, but nevertheless he held his tongue; for, thought he, would it indeed be worth while to prove to these fellows that it is their own stupidity which prevents them from appreciating Olimpia's profound and brilliant parts? One day Siegmund said to him, "Pray, brother, have the kindness to tell me how you, a sensible fellow, came to lose your head over that Miss Wax-face—that wooden doll across there?" Nathanael was about to fly into a rage, but he recollected himself and replied, "Tell me, Siegmund, how came it that Olimpia's divine charms could escape your eye, so keenly alive as it always is to beauty, and your acute perception as well? But Heaven be thanked for it, otherwise I should have had you for a rival, and then the blood of one of us would have had to be spilled." Siegmund, perceiving how matters stood with his friend, skilfully interposed and said, after remarking that all argument with one in love about the object of his affections was out of place, "Yet it's very strange that several of us have formed pretty much the same opinion about Olimpia. We think she is—you won't take it ill, brother?—that she is singularly statuesque and soulless. Her figure is regular, and so are her features, that can't be gainsaid; and if her eyes were not so utterly devoid of life, I may say, of the power of vision, she might pass for a beauty. She is strangely measured in her movements, they all seem as if they were dependent upon some wound-up clock-work. Her playing and singing has the disagreeably perfect, but insensitive time of a singing machine, and her dancing is the same. We felt quite afraid of this Olimpia, and did not like to have anything to do with her; she seemed to us to be only acting *like* a living creature, and as if there was some secret at the bottom of it all." Nathanael did not give way to the bitter feelings which threatened to master him at these words of Siegmund's; he fought down and got the better of his displeasure, and merely said, very earnestly, "You cold prosaic fellows may very well be afraid of her. It is only to its like that the poetically organised spirit unfolds itself. Upon me alone did her loving glances fall, and through my mind and thoughts alone did they radiate; and only in her love can I find my own self again. Perhaps, however, she doesn't do quite right not to jabber a lot of nonsense and stupid talk like other shallow people. It is true, she speaks but few words; but the few words she does speak are genuine hiero-glyphs of the inner world of Love and of the higher cognition of the intellec-tual life revealed in the intuition of the Eternal beyond the grave. But you have no understanding for all these things, and I am only wasting words."

"God be with you, brother," said Siegmund very gently, almost sadly, "but it seems to me that you are in a very bad way. You may rely upon me, if all— No, I can't say any more." It all at once dawned upon Nathanael that his cold prosaic friend Siegmund really and sincerely wished him well, and so he warmly shook his proffered hand.

Nathanael had completely forgotten that there was a Clara in the world, whom he had once loved—and his mother and Lothair. They had all vanished from his mind; he lived for Olimpia alone. He sat beside her every day for hours together, rhapsodising about his love and sympathy enkindled into life, and about psychic elective affinity—all of which Olimpia listened to with great reverence. He fished up from the very bottom of his desk all the things that he had ever written—poems, fancy sketches, visions, romances, tales, and the heap was increased daily with all kinds of aimless sonnets, stanzas, canzo-nets. All these he read to Olimpia hour after hour without growing tired; but then he had never had such an exemplary listener. She neither embroidered, nor knitted; she did not look out of the window, or feed a bird, or play with a little pet dog or a favourite cat, neither did she twist a piece of paper or any-thing of that kind round her finger; she did not forcibly convert a yawn into a low affected cough—in short, she sat hour after hour with her eyes bent unchangeably upon her lover's face, without moving or altering her position, and her gaze grew more ardent and more ardent still. And it was only when at last Nathanael rose and kissed her lips or her hand that she said, "Ach! Ach!" and then "Good-night, dear." Arrived in his own room, Nathanael would break out with, "Oh! what a brilliant—what a profound mind! Only you— you alone understand me." And his heart trembled with rapture when he reflected upon the wondrous harmony which daily revealed itself between his own and his Olimpia's character; for he fancied that she had expressed in respect to his works and his poetic genius the identical sentiments which he himself cherished deep down in his own heart in respect to the same, and even as if it was his own heart's voice speaking to him. And it must indeed have been so; for Olimpia never uttered any other words than those already mentioned. And when Nathanael himself in his clear and sober moments, as, for instance, directly after waking in a morning, thought about her utter passivity and taciturnity, he only said, "What are words—but words? The glance of her heavenly eyes says more than any tongue of earth. And how can, anyway, a child of heaven accustom herself to the narrow circle which the exigencies of a wretched mundane life demand?"

Professor Spalanzani appeared to be greatly pleased at the intimacy that had sprung up between his daughter Olimpia and Nathanael, and showed the young man many unmistakable proofs of his good feeling towards him; and when Nathanael ventured at length to hint very delicately at an alliance with Olimpia, the Professor smiled all over his face at once, and said he should allow his daughter to make a perfectly free choice. Encouraged by these words, and with the fire of desire burning in his heart, Nathanael resolved the very next day to implore Olimpia to tell him frankly, in plain words, what he had long read in her sweet loving glances,—that she would be his for ever. He looked for the ring which his mother had given him at parting; he would present it to Olimpia as a symbol of his devotion, and of the happy life he was to lead with her from that time onwards. Whilst looking for it he came across his letters from Clara and Lothair; he threw them carelessly aside, found the ring, put it in his pocket, and ran across to Olimpia. Whilst still on the stairs, in the entrance-passage, he heard an extraordinary hubbub; the noise seemed to proceed from Spalanzani's study. There was a stamping—a rattling—pushing—knocking against the door, with curses and oaths intermingled. "Leave hold—leave hold—you monster—you rascal—staked your life and honour upon it?—Ha! ha! ha! ha!—That was not our wager—I, I made the eyes—I the clock-work.—Go to the devil with your clock-work—you damned dog of a watch-maker—be off—Satan—stop—you paltry turner—you infernal beast!—stop—begone—let me go." The voices which were thus making all this racket and rumpus were those of Spalanzani and the fearsome Coppelius. Nathanael rushed in, impelled by some nameless dread. The Professor was grasping a female figure by the shoulders, the Italian Coppola held her by the feet; and they were pulling and dragging each other backwards and forwards, fighting furiously to get possession of her. Nathanael recoiled with horror on recognising that the figure was Olimpia. Boiling with rage, he was about to tear his beloved from the grasp of the madmen, when Coppola by an extraordinary exertion of strength twisted the figure out of the Professor's hands and gave him such a terrible blow with her, that he reeled backwards and fell over the table all amongst the phials and retorts, the bottles and glass cylinders, which covered it: all these things were smashed into a thousand pieces. But Coppola threw the figure across his shoulder, and, laughing shrilly and horribly, ran hastily down the stairs, the figure's ugly feet hanging down and banging and rattling like wood against the steps. Nathanael was stupefied;— he had seen only too distinctly that in Olimpia's pallid waxed face there were no eyes, merely black holes in their stead; she was an inanimate puppet. Spalanzani was rolling on the floor; the pieces of glass had cut his head and breast and arm; the blood was escaping from him in streams. But he gathered his strength together by an effort.

"After him—after him! What do you stand staring there for? Coppelius— Coppelius—he's stolen my best automaton—at which I've worked for twenty years—staked my life upon it—the clock-work—speech—movement— mine—your eyes—stolen your eyes—damn him—curse him—after him— fetch me back Olimpia—there are the eyes." And now Nathanael saw a pair of bloody eyes lying on the floor staring at him; Spalanzani seized them with his uninjured hand and threw them at him, so that they hit his breast. Then madness dug her burning talons into him and swept down into his heart, rending his mind and thoughts to shreds. "Aha! aha! aha! Fire-wheel—fire-wheel! Spin round, fire-wheel! merrily, merrily! Aha! wooden doll! spin round, pretty wooden doll!" and he threw himself upon the Professor, clutching him fast by the throat. He would certainly have strangled him had not several people, attracted by the noise, rushed in and torn away the madman; and so they saved the Professor, whose wounds were immediately dressed. Siegmund, with all his strength, was not able to subdue the frantic lunatic, who continued to scream in a dreadful way, "Spin round, wooden doll!" and to strike out right and left with his doubled fists. At length the united strength of several succeeded in overpowering him by throwing him on the floor and binding him. His cries passed into a brutish bellow that was awful to hear; and thus raging with the harrowing violence of madness, he was taken away to the madhouse.

Before continuing my narration of what happened further to the unfortunate Nathanael, I will tell you, indulgent reader, in case you take any interest in that skilful mechanician and fabricator of automata, Spalanzani, that he recovered completely from his wounds. He had, however, to leave the university, for Nathanael's fate had created a great sensation; and the opinion was pretty generally expressed that it was an imposture altogether unpardonable to have smuggled a wooden puppet instead of a living person into intelligent tea-circles,—for Olimpia had been present at several with success. Lawyers called it a cunning piece of knavery, and all the harder to punish since it was directed against the public; and it had been so craftily contrived that it had escaped unobserved by all except a few preternaturally acute students, although everybody was very wise now and remembered to have thought of several facts which occurred to them as suspicious. But these latter could not succeed in making out any sort of a consistent tale. For was it, for instance, a thing likely to occur to any one as suspicious that, according to the declaration of an elegant beau of these tea-parties, Olimpia had, contrary to all good manners, sneezed oftener than she had yawned? The former must have been, in the opinion of this elegant gentleman, the winding up of the concealed clockwork; it had always been accompanied by an observable creaking, and so on. The Professor of Poetry and Eloquence took a pinch of snuff, and, slapping the lid to and clearing his throat, said solemnly, "My most honourable ladies and gentlemen, don't you see then where the rub is? The whole thing is an allegory, a continuous metaphor. You understand me? *Sapienti sat.*" But several most honourable gentlemen did not rest satisfied with this explanation; the history of this automaton had sunk deeply into their souls, and an absurd mistrust of human figures began to prevail. Several lovers, in order to be fully convinced that they were not paying court to a wooden puppet, required that their mistress should sing and dance a little out of time, should embroider or knit or play with her little pug, &c., when being read to, but above all things else that she should do something more than merely listen—that she should frequently speak in such a way as to really show that her words presupposed as a condition some thinking and feeling. The bonds of love were in many cases drawn closer in consequence, and so of course became more engaging; in other instances they gradually relaxed and fell away. "I cannot really be made responsible for it," was the remark of more than one young gallant. At the tea-gatherings everybody, in order to ward off suspicion, yawned to an incredible extent and never sneezed. Spalanzani was obliged, as has been said, to leave the place in order to escape a criminal charge of having fraudulently imposed an automaton upon human society. Coppola, too, had also disappeared.

When Nathanael awoke he felt as if he had been oppressed by a terrible nightmare; he opened his eyes and experienced an indescribable sensation of mental comfort, whilst a soft and most beautiful sensation of warmth pervaded his body. He lay on his own bed in his own room at home; Clara was bending over him, and at a little distance stood his mother and Lothair. "At last, at last, O my darling Nathanael; now we have you again; now you are cured of your grievous illness, now you are mine again." And Clara's words came from the depths of her heart; and she clasped him in her arms. The bright scalding tears streamed from his eyes, he was so overcome with mingled feelings of sorrow and delight; and he gasped forth, "My Clara, my Clara!" Siegmund, who had staunchly stood by his friend in his hour of need, now came into the room. Nathanael gave him his hand—"My faithful brother, you have not deserted me." Every trace of insanity had left him, and in the tender hands of his mother and his beloved, and his friends, he quickly recovered his strength again. Good fortune had in the meantime visited the house; a niggardly old uncle, from whom they had never expected to get anything, had died, and left Nathanael's mother not only a considerable fortune, but also a small estate, pleasantly situated not far from the town. There they resolved to go and live, Nathanael and his mother, and Clara, to whom he was now to be married, and Lothair. Nathanael was become gentler and more childlike than he had ever been before, and now began really to understand Clara's supremely pure and noble character. None of them ever reminded him, even in the remotest degree, of the past. But when Siegmund took leave of him, he said, "By heaven, brother! I was in a bad way, but an angel came just at the right moment and led me back upon the path of light. Yes, it was Clara." Siegmund would not let him speak further, fearing lest the painful recollections of the past might arise too vividly and too intensely in his mind.

The time came for the four happy people to move to their little property. At noon they were going through the streets. After making several purchases they found that the lofty tower of the town-house was throwing its giant shadows across the market-place. "Come," said Clara, "let us go up to the top once more and have a look at the distant hills." No sooner said than done. Both of them, Nathanael and Clara, went up the tower; their mother, however, went on with the servant-girl to her new home, and Lothair, not feeling inclined to climb up all the many steps, waited below. There the two lovers stood arm-in-arm on the topmost gallery of the tower, and gazed out into the sweet-scented wooded landscape, beyond which the blue hills rose up like a giant's city.

"Oh! do look at that strange little grey bush, it looks as if it were actually walking towards us," said Clara. Mechanically he put his hand into his sidepocket; he found Coppola's perspective and looked for the bush; Clara stood in front of the glass. Then a convulsive thrill shot through his pulse and veins;

pale as a corpse, he fixed his staring eyes upon her; but soon they began to roll, and a fiery current flashed and sparkled in them, and he yelled fearfully, like a hunted animal. Leaping up high in the air and laughing horribly at the same time, he began to shout, in a piercing voice, "Spin round, wooden doll! Spin round, wooden doll!" With the strength of a giant he laid hold upon Clara and tried to hurl her over, but in an agony of despair she clutched fast hold of the railing that went round the gallery. Lothair heard the madman raging and Clara's scream of terror: a fearful presentiment flashed across his mind. He ran up the steps; the door of the second flight was locked. Clara's scream for help rang out more loudly. Mad with rage and fear, he threw himself against the door, which at length gave way. Clara's cries were growing fainter and fainter,— "Help! save me! save me!" and her voice died away in the air. "She is killed— murdered by that madman," shouted Lothair. The door to the gallery was also locked. Despair gave him the strength of a giant; he burst the door off its hinges. Good God! there was Clara in the grasp of the madman Nathanael, hanging over the gallery in the air; she only held to the iron bar with one hand. Quick as lightning, Lothair seized his sister and pulled her back, at the same time dealing the madman a blow in the face with his doubled fist, which sent him reeling backwards, forcing him to let go his victim.

Lothair ran down with his insensible sister in his arms. She was saved. But Nathanael ran round and round the gallery, leaping up in the air and shouting, "Spin round, fire-wheel! Spin round, fire-wheel!" The people heard the wild shouting, and a crowd began to gather. In the midst of them towered the advocate Coppelius, like a giant; he had only just arrived in the town, and had gone straight to the market-place. Some were going up to overpower and take charge of the madman, but Coppelius laughed and said, "Ha! ha! wait a bit; he'll come down of his own accord;" and he stood gazing upwards along with the rest. All at once Nathanael stopped as if spell-bound; he bent down over the railing, and perceived Coppelius. With a piercing scream, "Ha! foine oyes! foine oyes!" he leapt over.

When Nathanael lay on the stone pavement with a broken head, Coppelius had disappeared in the crush and confusion.

Several years afterwards it was reported that, outside the door of a pretty country house in a remote district, Clara had been seen sitting hand in hand with a pleasant gentleman, whilst two bright boys were playing at her feet. From this it may be concluded that she eventually found that quiet domestic happiness which her cheerful, blithesome character required, and which

Nathanael, with his tempest-tossed soul, could never have been able to give her.

Ernst Theodor Wilhelm Hoffmann (1776–1822) was a Prussian polymath—a jurist, caricaturist, composer, music critic, draftsman… and one of the major authors of the Romantic movement. Three of his stories formed the basis for Offenbach's famous opera *The Tales of Hoffmann* (1882). Hoffmann (who is usually known as E.T.A. Hoffmann, the "A" standing for Amadeus) wrote some of his best known stories at around the same time that Mary Shelley was writing the first science fiction novel—*Frankenstein*. These authors were deeply engaged with the scientific debates then raging.

Commentary

"The Sandman" is the oldest tale in this book. Hoffman published it two centuries ago, in 1817, in a collection entitled *Nachtstücke* (or *Night Pieces*). The writing shows its age, and the protagonist Nathaniel is unlikely to appeal to modern readers. And yet of all the stories in this book "The Sandman" touches most directly on contemporary concerns.

In Hoffman's tale, Nathaniel falls in love with a beautifully constructed automaton—what we would now call an android, a robot with a human form. It's unclear whether Nathaniel's infatuation with Olimpia is merely evidence of the insanity that ultimately leads to his suicide or whether it exacerbates his unstable mental state. The reaction of the other characters to Olimpia is, I believe, more interesting. Although they sense something strange and unlikeable about her—she is the antithesis of Nathaniel's fiancée, the intelligent and practically minded Clara—they don't immediately identify her as an automaton. When the truth about Olimpia comes out, however, people start doubting their own powers of discrimination. Hoffman thus raises numerous profound questions. Can one fall in love with an android? What sorts of human–android relationships will society find permissible? Will we always be able to distinguish human from android?

Asimov, in his famous series of positronic robot stories, examined such questions from a variety of angles. Asimov, however, was an SF writer attempting to imagine the distant future. Even the most technologically optimistic among us would agree we are still decades away from developing an android with the same wisdom and charisma as Asimov's most successful robot, Daneel Olivaw. So at first glance it might seem reasonable to categorise these questions as being of interest only to the readers and writers of SF. But that would be a mistake. Our whole society should be grappling with them *now*.

* * *

Engineers can't yet manufacture a true android—a robot that looks, moves, and sounds human. AI experts can't program a robot with the general intelligence necessary to navigate the world in all its messy complications. Psychologists can't define the quality of "humanness"—a property we all recognize in each other—let alone tell engineers how to imbue a machine with this quality. With the current state-of-the-art, devices are either clearly robots (acting efficiently in one particular niche, but with little attempt made to make them humaniform) or else clumsy constructs that might possess the surface appearance of a person but are not much more sophisticated than a waxwork dummy. In either case, most humans immediately know they are looking at non-humans. But advances in engineering, AI, and psychology have already enabled the construction of machines—androids or robots— that are "good enough" for some people to form an attachment to them.

Our evolutionary history has hard-wired our brains into attributing agency to objects that move in particular ways. This makes perfect sense. None of our ancestors would have gained an advantage by trying to second-guess the motivations of a rock rolling down a mountainside, but they *would* have found it useful to apprehend the mental state of an approaching tiger. This emotional hard-wiring is so powerfully ingrained that some people can form an attachment to *anything*. Consider, for example, an experiment carried out in 2011 by the computer scientists John Harris and Ehud Sharlin.

Harris and Sharlin sat their human subjects in a room, with only a stick of balsa wood for company. The stick was in fact a simple robot: it was attached to a gearing mechanism through which a human operator (hidden from view) could use a joystick to make the balsa wood move in certain ways. Now, you'd think people would call a stick a stick. But a majority of subjects chose to attribute agency to the stick: they perceived it as possessing goals and internal thought processes. They tried to predict what the stick was "thinking" based on how it moved, just as our ancestors might have attributed a state of mind to a passing wild animal.

This experiment demonstrates that we shouldn't be surprised if people start forming attachments to androids. The androids don't need to be a perfect facsimile of the human form in order for emotional attachment to occur. After all, the androids will appear far more human than does a stick of balsa wood. Indeed, people are *already* forming attachments to less-than-perfect androids. Consider, for example, the Telenoid (see Fig. 6.1).

Fig. 6.1   A collection of early Telenoids (Credit: Osamu Iwasaki)

The Japanese roboticist Hiroshi Ishiguro developed the Telenoid—an 80 cm, 5 kg droid made of silicone rubber—in 2010. The Telenoid has shortened, stubby arms and, instead of legs, a rounded stump; the skin-colored surface is perfectly smooth; deep-set, jet-black eyes stare out from a face that's calm and untroubled. Personally, I find the Telenoid creepy. But that's not how dementia sufferers react.

For elderly dementia sufferers, a Telenoid is something to be cradled, rocked, cared for. This gives them some measure of peace. Presumably these patients don't appreciate how the Telenoid's movement is controlled by a human using teleoperation, nor that what they hold in their hands is rubber rather than flesh. Caregivers describe how their patients lavish affection on a Telenoid. Indeed, those caregivers have discovered they don't need to provide patients with a sort-of-humanoid baby in order to see the same beneficial effect: the Paro therapeutic robot looks like a fluffy white seal and many dementia sufferers treat it with as much affection as they would a real pet.

In both these cases the innate human capacity for empathy is being purposefully manipulated—but does this matter? Is it of any importance *how* a Telenoid pseudo-baby or a Paro fake seal gives people comfort, so long as those people feel happier? Or should we be concerned about the ethics of outsourcing such an important task—providing emotional support to human beings—to robots?

Robots such as the Telenoid or the Paro have specialist applications in care settings, but most members of the general public are unlikely to find these devices more than momentarily diverting. The problem is these robots are *stupid*. Although they are effective at eliciting an emotional response they lack the general intelligence necessary to keep our attention. If a youngster received one for a Christmas present then by Boxing Day the device's limited repertoire of sayings and behaviours would be exhausted; by New Year those sayings and behaviours would have changed from "cute" to "irritating"; by Easter the device would be gathering dust. Even the much-hyped *Sophia* humanoid (see Fig. 6.2), which since its activation in 2015 has received the sort of media attention usually reserved for human celebrities, is essentially just a mobile chatbot. But that's the state of the art *now*. As we saw in the previous chapter, AI is advancing rapidly. At some point—a few decades from now, perhaps—a program capable of conversing meaningfully with people will be running inside an android capable of eliciting empathy while navigating the messy, ever-changing, everyday world. Such an android wouldn't be like Hoffman's Olimpia; it would be much more like Asimov's Daneel Olivaw.

Fig. 6.2 Sophia—a humanoid said to be modeled upon Audrey Hepburn (though personally I don't see the resemblance). Sophia can mimic certain human gestures and can converse simply on a limited number of predefined topics (Credit: ITU Pictures)

These androids of the future could be constructed not only to negotiate the human world but also to be better looking than humans. (Different people of course define "better looking" in different ways. But that won't matter: engineers will be able to manufacture an android to align with any particular person's definition of "beautiful".) An android's skin can be made purer than a human's skin; an android won't suffer from disease; an android won't get tired. An android could be designed to be extremely attractive to you. If such an android told you "I love you"—how would you react?

From our vantage point, early in the 21st century, the thought of someone entering into a serious relationship with an inanimate object is likely to produce sniggers. At present, the so-called "sexbots" some people fret about are hardly more advanced than blow-up rubber dolls. But that's the situation *now*. By the middle of the century there might be androids specifically manufactured with attractiveness and beauty in mind, androids capable of meaningful conversation and interaction. Occurrences of human–android relationships would be inevitable—it goes against what we know of human nature to suppose human–android relationships *won't* develop.

The questions raised by such relationships are legion. How would such relationships change human society? Might people prefer to partner with perfect androids rather than with imperfect humans? And if an android *acted* as if it had feelings and *said* it had feelings would we accept it *did* have feelings? (After all, the Turing test suggests that if a robot acts as if it were thinking then we should accept it is thinking. Shouldn't we grant the same when it comes to expressing emotions?) And if we argue that androids feel things merely because they are programmed so to do, does it follow that human emotions are also merely a matter of programming? Human programming might involve hormones and neurons rather than lines of code and transistors, but does that make a difference? And if we say it *doesn't* make a difference, must we then change how we view ourselves—are humans and androids nothing more than computers, the former based on carbon, the latter based on silicon? In that case, should androids themselves have a say in all of this?

* * *

Isaac Asimov thought about these questions perhaps more broadly than anyone else in the 20th century. In his novels and short stories he examined numerous possible answers. Two of his robot novels from the 1950s presented two opposing views of how humankind might react to technically advanced androids. In *The Caves of Steel* Asimov imagined a future Earth on which people have chosen to ban all use of robots. Billions of humans are crowded together, living in their caves of steel, coping as best they can with resources made scarce by overpopulation. In this future, technophobia wins out—and the result is not pleasant (although for the people living in this society it of course seems perfectly natural). His companion novel *The Naked Sun* is set in the same time period, but the action is located on the planet Solaria. Solarian society is in many ways the polar opposite of Earth's: the human population is strictly controlled at 20,000 but for every person there are ten thousand robots. People are taught to avoid personal contact. They live alone on vast estates and choose to "view" rather than meet one another. For these people the issue of procreation is thoroughly embarrassing, and babies are born in "birthing centres". This is a future in which technophilia wins out—and the result is equally unpleasant (although again for the people living in this society it all seems perfectly natural). Elements of either of these futures seem plausible. If robots cause mass unemployment and trigger social changes the majority of people find offensive then perhaps their use will be curtailed. If androids turn out to be pleasant company, and a useful addition to our society, then perhaps they'll become our helpmates.

Asimov foresaw other possibilities, though. Perhaps the future belongs neither to a humans-only world nor one dominated by androids but to a world in which humans and androids merge. A few individuals have already taken tentative steps towards such a future. People have had implants fixed to enable them to "hear" colour or "feel" earthquakes; they have fitted prosthetic arms and legs to improve their movement; they have inserted subcutaneous biometric chips so their body status (blood pressure, stress level, and so on) is transmitted to their internet-connected environment, which can be programmed to respond appropriately. Could it be that advances in digital technology, combined with the advances in genetic technology we discussed in earlier chapters, lead not to androids with their silicone perfection but to cyborgs with bizarre and outlandish forms?

* * *

The options outlined above belong to a future many of us won't live to see. Right now, however, it's already possible to discern the ever-increasing influence of digital technology on our society. As humans move to an increasingly internet-connected world we are producing vast amounts of "digital exhaust"—vast data sets that our current, clumsy versions of AI are sifting through and thereby producing useful information in a way that's impossible for human workers. Androids and cyborgs are for tomorrow. Big data is changing the world today. And big data, in a sense, is the theme of Chapter 7.

Notes and Further Reading

a collection entitled Nachtstücke—Hoffman's story "The Sandman" was first published in 1815, and then appeared as the opening story of his two-volume collection *Nachtstücke* (Hoffman 1816–1817). *Nachtstücke* contained eight stories, but it is "The Sandman" that has attracted most attention. This might be in part because Freud (1919) published an influential essay interpreting some of the motifs of Hoffman's story in terms of his own psychoanalytical framework.

famous series of positronic robot stories—As mentioned in the previous chapter, most of Asimov's robot-themed short stories were anthologized in Asimov (1982).

"humanness"—a property we all recognize in each other—Actually, we don't all always recognize humanness. Those suffering from Capgras delusion are convinced someone close to them has been replaced by an identical impostor. This is an extremely rare syndrome, but there are recorded cases in which a person with the delusion believes the replacement is a robot or an android. One young man with the syndrome became convinced his father had been replaced by a robot. He cut off his father's head and demanded the authorities check it for hidden electronics. See for example de Pauw et al. (2008).

the computer scientists John Harris and Ehud Sharlin—See Harris and Sharlin (2011).

Japanese roboticist Hiroshi Ishiguro developed the Telenoid—See, for example, Guizzo (2010).

the Paro therapeutic robot—For further details, see parorobots (n.d.).

Isaac Asimov thought about these questions—For the two novels mentioned in the text, see Asimov (1954, 1957). These novels are classics in the field, but they are now more than six decades old. Human–robot interactions have appeared in more recent novels, including one published in this very series: *The Hunter* (Genta 2014) involves "RGs"—robogirls and roboguys. One character in the novel describes an RG as being "seventy kilos of machinery, oved by electrohydraulic actuators controlled by a computer and covered with silicon flesh and synthetic skin". Genta foresees that "their use is the most private type of personal service: sex".

People have had implants—For example, Neil Harbisson had a wifi antenna implanted in his skull; the antenna translates EM radiation into audible vibrations. Moon Ribas had a seismic sensor implanted into her elbow. In 2010, these two artists co-founded the Cyborg Foundation (n.d.).

Bibliography

Asimov, A.: The Caves of Steel. Doubleday, New York (1954)

Asimov, A.: The Naked Sun. Doubleday, New York (1957)

Asimov, A.: The Complete Robot. Doubleday, New York (1982)

Cyborg Foundation: Home page. https://www.cyborgarts.com/ (n.d.)

de Pauw, K., Workman, L., Young, A.: The interview. Psychologist. **21**(Feb), 120–121 (2008)

Freud, S.: Das Unheimliche. Imago. **5**(5–6), 297–324 (1919)

Genta, G.: The Hunter. Springer, Berlin (2014)

Guizzo, E.: Telenoid R1: Hiroshi Ishiguro's newest and strangest android. IEEE Spectr. https://spectrum.ieee.org/automaton/robotics/humanoids/telenoid-r1-hiroshi-ishiguro-newest-and-strangest-android (2010)

Harris, J., Sahrlin, E.: Exploring the affect of abstract motion in social human–robot interaction. In: 20th IEEE International Symposium on Robot and Human Interactive Communication (2011)

Hoffman, E.T.A.: Nachtstücke. Realschulbuchhandlung, Berlin (1816–1817)

parorobots.com: Paro Therapeutic Robot. http://www.parorobots.com/ (n.d.)

7

Big Data

The Universal Library (Kurd Lasswitz)

(Translated from the Original by Heike and Stephen Webb)

"Come and take a seat here, Max," Professor Wallhausen said. "There's really nothing in my papers that would be of interest for your magazine. What can I get you—wine or beer?"

Max Burkel strolled over to the living-room table and raised his eyebrows thoughtfully. He let his not inconsiderable bulk sink into an armchair. "Actually, I'm teetotal now. But on holiday— ah, I see you have a splendid Kulmbacher there—thank you very much, dear girl—not so full! Well, cheers old chap! Cheers, my dear lady! Your good health, Miss Briggen! Ah, it's nice to be here again. But this isn't doing you any good, you know; you still have to write something for me."

"I really can't think of anything right now. In any case, there's so much superfluous stuff being written—and, unfortunately, printed—"

"You don't have to point that out to a harassed editor like me. But the question is: what is the superfluous stuff? Authors and readers take a quite different view of the matter. And whatever the book reviewers judge superfluous is exactly what reaches the likes of us. But I don't have to worry about that right now." Burkel rubbed his hands together with glee. "For the next three weeks my deputy editor can do the worrying for me."

© Springer Nature Switzerland AG 2019
S. Webb, *New Light Through Old Windows: Exploring Contemporary Science Through 12 Classic Science Fiction Tales*, Science and Fiction,
https://doi.org/10.1007/978-3-030-03195-4_7

"I sometimes wonder," Mrs. Wallhausen said, "how you can still find anything new to print. I'd have thought you must have tried out more or less everything that can be expressed with a few letters."

"True, you'd think so," Burkel replied, "but the human mind is inexhaustible—"

"In repetition, you mean."

"Thank God, yes," Burkel laughed. "But in terms of novelty, also."

"And yet," Professor Wallhausen observed, "one is able to use letters to represent everything ever experienced by humanity—whether that's historical happenings, scientific knowledge, poetic imagination, or the teachings of wisdom. Insofar as these things can be expressed using language. Our books, after all, preserve and disseminate the accumulated treasure resulting from human thought. However, the number of different combinations of a given number of letters is finite. It follows, therefore, that all possible literature must be printable in a finite number of volumes."

"Old friend," said Burkel, "you're talking as a mathematician rather than as a philosopher. How can the inexhaustible be finite?"

"If you'll indulge me, I can quickly work it out for you—the number of volumes there'll be in the Universal Library."

"Oh, uncle, will this be very academic?" Susanne Briggen asked.

"But Suse, for a young lady who has just come from a boarding school, surely nothing is too academic?"

"Thank you, uncle, but I just asked because I wondered whether I should bring my sewing because—because it helps me concentrate, you know?"

"Cleverclogs. You really just wanted to know whether I was going to drone on for hours! Probably not, in fact. But if you could just get me a sheet of paper and a pencil from my desk ..."

"Bring a table of logarithms too," Burkel added dryly.

"Oh for God's sake," Mrs Wallhausen protested.

"No, no, that's not necessary," the professor called. "And you don't need to show off your sewing either, Suse."

"Here's some more comfortable work," said Mrs Wallhausen, and she pushed a bowl of apples and nuts towards Susanne.

"Thanks," said Susanne, picking up the nutcracker. "Now I have something to crack your toughest problems."

"Let our friend have the first word," the professor began. "Tell me: if we dispense with different fonts and typefaces, and assuming we write only for the reader who's not fussy and is interested only in meaning, then ..."

Burkel interrupted. "There is no such reader."

"Just suppose there is. How many different characters would we need to print all texts, from great literature to light fiction?"

"Well," said Burkel, "if we confine ourselves to the uppercase and lowercase letters of the Latin alphabet, the usual punctuation marks, numerals, and— let's not forget—the space …"

Susanne glanced up inquisitively from her bowl of nuts.

"That's the piece of type a typesetter uses to separate individual words and to fill in blanks. So we wouldn't need too many characters. But for scientific and technical books! You mathematicians use a mountain of symbols."

"We can help ourselves here through the use indices—small numbers set above or below letters of the alphabet, a_0, a_1, a_2 and so on. That way we'd just need a second and third row of numbers from 0 to 9. In fact, with agreement, we could even use this system to represent words in any foreign text."

"Whatever. Well if your hypothetical ideal reader will put up with that too then we could probably express every conceivable thing with, say, one hundred different characters."

"So. Next question. How big should each volume be?"

"I believe one can pretty much exhaust a topic within five hundred pages. Let's say that there are 40 lines per page and 50 characters per line, where of course we are including punctuation marks, spaces, and so on. So we'll have $40 \times 50 \times 500$ characters per volume, which is—you do the math."

"One million," said the professor. "If, therefore, we take our 100 characters and put them together, in any order, enough times to fill a volume that has room for one million of them—well, we'll produce a piece of literature of some kind. And if we produce all possible combinations, purely mechanically, then we'll have all the works of literature that have already been written and all that can ever be written in the future."

Burkel slapped his friend powerfully on the shoulder. "I'm going to subscribe to your Universal Library! All future issues of my magazine will be ready-made. I won't have to bother any more about manuscripts. This is magnificent news for the publisher: the elimination of the author from the whole publishing business! The replacement of the writer by the combinatoric printing press! A technological triumph!"

"What?" cried Mrs Wallhausen. "*Everything* is in that library? The complete works of Goethe? The Bible? All the writings of all the philosophers who have ever lived?"

"Indeed. Along with all the interpretations that nobody has yet considered. You'll also find the lost works of Plato and Tacitus, along with their translations into all languages. Furthermore, the library contains all of my and my friend Burkel's future works; all forgotten and still-to-be delivered speeches to parliament; the universal declaration of peace and the history of the subsequent wars."

"And the book with the train timetables, uncle!" cried Susanne. "That's your favourite!"

"Of course. And the library will also contain all your essays for Miss Grazelau's German class."

"Oh, I wish I'd had that book when I was at boarding school. But I thought we were talking about filled volumes …"

"If I may, Miss Briggen," Burkel interrupted, "don't forget the space character. Even the smallest verse could have a volume for itself, with all other pages empty. On the other hand, even the longest work can be accommodated: if it doesn't fit into a single volume it can be continued through several."

"I wouldn't fancy having to find something in the library," said Mrs Wallhausen.

"That's the catch," Professor Wallhausen said, smiling. He leaded back in his chair and watched the smoke curl up from his cigar. "You might think searching would be made easier by the fact the library necessarily contains its own catalogue—"

"Good, surely!"

"Yes, but how will you find it? And if you were to find an index volume it wouldn't help you because you couldn't be sure it wasn't one of those in which the library is indexed in every possible misleading manner."

"Damn! That's true."

"Hmm! There are a number of other little difficulties. For example, suppose we take the first volume in our library. Its first page is empty, likewise the second, and so on, all 500 pages. This is the volume in which the space character has been repeated one million times."

"At least it won't contain any nonsense," Mrs. Wallhausen remarked.

"A consolation, I suppose. Now let's take the second volume. Empty, completely empty, until the last page, in the millionth position, there appears a shy letter 'a'. Same thing in the third volume, but the 'a' has moved up one place and the final character is once again a space. And so with each volume of the first million volumes the 'a' moves up, place by place, until it reaches first place in the first volume of the second million. There's nothing else of interest in this volume. Things continue this way through the first hundred million volumes, until each of the hundred individual characters has made its lonely way from last place to first place. The same thing then happens with the combination 'aa', or with any other two characters, in all possible positions. One volume contains only full stops, another only question marks."

"Well," Burkel said, "it should be easy enough to recognize those meaningless volumes and discard them."

"Perhaps. But the worst is yet to come. Suppose you've found a volume that seems to make sense. Suppose, for example, you want to look something up in *Faust* and you manage to find a volume with the correct beginning. But after you've read a few pages it suddenly continues 'Papperle, happerle, nothing is there!' or simply 'aaaaa …' Or you find a table of logarithms but you don't know whether the numbers in it are right, since our library contains not only everything that's correct but also everything that's incorrect. You must take care not to be misled by chapter headings. A volume might begin 'History of the Thirty Years War' and then continue: 'When Prince Blücher married the Queen of Dahomey at Thermopylae …'"

"Oh, uncle, that's perfect for me!" laughed Susanne. "The books I could write—I'm developing quite a talent for mixing things up. The library would definitely contain the opening lines of *Iphigenia in Tauris*, as I once recited them:

'Out in your shadows, brisk treetops,
Obeying need, not my own desires,
I want to sit down on this stone bench.'

If the play had been printed like that I'd have been vindicated. And I'd also be able to find the long letter I wrote to the two of you. It went missing before I could find it. Mika put her schoolbooks down on top of it. Oh dear." Susanne broke off in embarrassment, and brushed her unruly brown hair from her brow. "Miss Grazelau expressly told me that I should take care not to babble so much!"

"Here you are quite justified," her uncle consoled her. "In our library you'll find not only all your letters but also all the speeches you've made or will ever make."

"Oh, then best not to tell anyone about this library!"

"Don't worry. Those letters aren't signed just by you, but also signed by Goethe, and indeed by all possible names. Our friend Burkel, for example, would find his signature attached to so many articles of every possible kind of libel that he wouldn't live long enough to serve the prison sentences. There'll be a book by him where it states after every sentence that the sentence is false; and another book containing the identical sentences except that everything is sworn to be true …"

"Enough," laughed Burkel. "I knew straight away you'd be spinning us a yarn. So I'm cancelling my subscription to the Universal Library. It would be impossible to separate sense from nonsense, truth from falsehood. If I find millions of volumes all claiming to be the true history of Germany during the twentieth century, and all of them contradicting each other, I'd be better off reading the original works of the historians. I'll give it a miss."

"Very wise! Otherwise you'd have taken on a huge burden. But I wasn't 'spinning a yarn', Burkel. I didn't claim you could *use* the Universal Library in practice. Rather, I merely observed that it's possible to calculate exactly how many volumes would be required for a Universal Library to contain all possible literature, the meaningful as well as the nonsensical."

"Then go ahead and calculate," Mrs Wallhausen sighed. "I can see that this blank sheet of paper bothers you."

"Oh, it's simple enough to do in my head," said the professor. "All we need to do is to understand how the library will be produced. First, we set down each one of our hundred characters. Then, to each of the hundred characters, we add every one of our hundred characters—so we have one hundred times one hundred groups of two characters each. Upon adding the third set of our hundred characters we get $100 \times 100 \times 100$ groups of three characters each, and so on. And since we have one million possible positions per volume, we'd have as many volumes as the number we get when we multiply 100 by itself a million times. Now 100 is the same as ten times ten, so we obtain the same figure if we multiply 10 by itself two million times. This is just a one followed by two million zeros. Here is the answer, then—ten to the power of two million."

The professor wrote $10^{2,000,000}$ and held up the paper.

"You make things too easy for yourself," Mrs Wallhausen said. "Write out the number."

"I'll do nothing of the kind. It would take me at least two weeks, night and day, without any breaks. If you printed that number, it would be about four kilometers long."

"Wow," said Susanne. "What's the number called?"

"It doesn't have a name. The number is so colossal there's no way to grasp its magnitude, even though it's finite. Whatever huge quantities you care to mention, they'd disappear compared to this numerical monster."

"How about expressing it in trillions?" Burkel asked.

"Well, a trillion is a nice number, a million million, a one followed by 18 zeros. Divide the number of our volumes by a trillion and you cancel 18 of the two million zeros. You get a number with 1,999,982 zeros, which is just as ungraspable. But wait a moment ..." The professor began scribbling a few numbers on the paper.

"I *knew* it!" his wife said. "I knew there'd be calculations."

"Already finished. Do you realise what this number means for our library? Suppose each of our volumes is two centimeters thick and we place all of them in a single row. How long do you think this row would be?"

He looked round triumphantly, as the others remained silent. Suddenly Susanne said, "I know! Want me to tell you?"

"Go ahead, Suse."

"Twice as many centimeters as the number of volumes in the library."

"Bravo!" the others said. "That will do nicely."

"Yes," said the professor. "But let's examine this more closely. You all know that light travels 300,000 kilometers in one second, so in one year light travels about ten billion kilometers, which is the same as a trillion centimeters. So if a librarian could move at the speed of light it would still take him two years to pass just a trillion volumes. To go from one end of the library to the other at the speed of light would take twice as many years as there are trillions of volumes in the library. We've spoken about that number before: a one followed by 1,999,982 zeros. I'm trying to make clear that we have as little chance of imagining the number of years it takes light to travel across the library as we have of imagining the number of volumes in the library. And this makes it plain that the task of forming a mental image of this number is quite hopeless. Even though it's a *finite* number."

As the professor began to set the paper aside, Burkel said: "If the ladies will permit, I have one more question. I suspect you've calculated a library that the universe is too small to contain."

"We'll see in a moment," the professor answered, and he began calculating again. "Let's assume we pack the library into 1000-volume boxes, with each box having a capacity of one cubic meter. To contain the library would require the entire universe, out to the farthest visible galaxies, so many times over that the number of book-packed universes would have only about 60 zeros fewer than the number with two million zeros, which is our figure for the number of volumes. So the fact remains—there's no way we can get close to this gigantic number."

"You see," said Burkel, "I was right all along that it's inexhaustible."

"No, it's not. Just subtract it from itself and you have 'zero'. The number is finite, it is well defined as a concept. The only surprise is this: we can write down using just a few digits the number of volumes that comprise the seemingly infinite amount of all possible literature. But if we then try to visualize it—for example, if we try to imagine finding a specific volume in our universal library—we come to the realisation that we can't comprehend the result of a logical argument that we ourselves developed."

Burkel nodded earnestly. "Reason is infinitely greater than understanding."

"What does that riddle mean?" asked Mrs Wallhausen.

"I mean simply that we can reason correctly about infinitely more things than we can determine by experience. That logic is infinitely more powerful than the senses."

"That's rather uplifting," remarked Professor Wallhausen. "Sense experience decays over time, but a logical argument is independent of time. It is universally valid. And because logical arguments are nothing more than the thinking of humanity itself, we possess in this timeless good a share of the immutable laws of the divine, of the destiny of the infinite power of creation. The foundations of mathematics are based on this."

"Good," said Burkel, "the laws give us confidence in the truth. But we can only use them if we fill their ideal form with material from lived experience. In other words, when we have found the volume we need from the library."

Wallhausen agreed, and his wife softly quoted:

No mortal man
shall ever compete
with the gods.
If he rises
and touches
the stars with his head,
then his unsteady feet
will have nothing to stand on,
and the clouds and the winds
will play with him.

"The great poet captures it exactly," said Professor Wallhausen. "But without the laws of logic there'd be nothing to lift us to the stars and beyond. It's just that we mustn't leave the solid ground of experience. We shouldn't look in the Universal Library for the volume we require, but instead write it ourselves through committed, serious, honest work."

"Chance plays, reason creates," Burkel concluded. "That's why you'll write down tomorrow what you played today. That way I'll get my article to take away with me after all."

"I can do that for you," Wallhausen laughed. "But I'll tell you right now: your readers will conclude it's taken from one of the superfluous volumes! Suse—what are you doing?"

"I'm doing something reasonable," she said gravely. "I'm going to meet form with matter."

And she refilled the glasses.

Kurd Lasswitz (1848–1910) was a German philosopher and historian of science. He also authored a number of science fiction novels and short stories, for which he has been called the 'father of German SF'. The Kurd Lasswitz Preis is still presented annually to award the best German science fiction published in the previous year.

Note on the Translation

Kurd Lasswitz published this story in Germany in 1904. The German–American science writer Willy Ley published an English translation, which appeared in the popular 1958 book *Fantasia Mathematica*, edited by Clifton Fadiman. Ley's translation, however, focused strongly on the mathematics, and he chose to omit numerous elements of the story while adding some of his own material. I've therefore presented an original translation here, which I hope better retains the feel of a turn-of-the-century German story. (In this I have been guided by my wife Heike, who is a native German speaker.)

Some specific notes follow.

Characters—four characters appear in the story: Professor Wallhausen; his wife; their niece, Miss Susanne Briggen (whom the Wallhausens sometimes refer to by the diminutive "Suse"); and Wallhausen's friend, an editor, Max Burkel. In the original story, the professor's wife is referred to as "Frau Professor" and "die Hausfrau"; we've translated these throughout by the phrase Mrs Wallhausen. We aren't told the given name of either the professor or his wife.

Kulmbacher—a beer. The Kulmbacher brewery, based in a small town near Bayreuth, is still producing beer. I can strongly recommend it.

table of logarithms—a more modern translation would probably use "smartphone" rather than "table of logarithms" because smartphone apps can handle advanced mathematics. Before people had such technological wonders available to them they'd perform tedious mathematical calculations with the help of log tables.

the book with the train timetables—the original story talks about the *Reichskursbuch*, or Imperial timetables. This was a book of train schedules and ship and canal connections. The first such book was published in 1850.

look something up in Faust—Ley's translation, for an English-speaking audience, employed Shakespeare as the author whose name everyone would know. The closest equivalent in Germany is perhaps Goethe, and his play *Faust* is the best-known literary work in German.

begin 'History of the Thirty Years War'—Lasswitz deliberately mixes up various events to demonstrate why it's impossible to trust a history text taken from the Universal Library. The Thirty Years War (1618–1648) was a deadly conflict in Central Europe, which began as an altercation between Protestant and Catholic states and then escalated into more general war. Prince Blücher fought against Napoleon at the Battle of Waterloo, some two hundred years after the Thirty Years War began. The kingdom of Dahomey was annexed by the French in 1894; it's now Benin. It's unlikely Blücher would have ever met the Queen of Dahomey; and even he had, he wouldn't have married her at Thermopylae, which is best known for the battle that took place there in 480 BC during the second Persian invasion of Greece!

the opening lines of Iphigenia in Tauris, *as I once recited them*—When Susanne recited the opening lines of Goethe's play Iphigenia in Taurus she mixed up well known lines from three different German plays. "Out in your shadows, brisk treetops" is correct—it's the first line from Goethe's play; but "Obeying need, not my own desires" comes from Schiller's *The Bride of Messina*; and the line "I want to sit down on this stone bench" comes from Schiller's *Wilhelm Tell*. An English version of this passage might be from Hamlet's soliloquy: "To be or not to be, that is the question. To live is the rarest thing in the world; most people exist, that is all. What's breaking into a bank compared with founding a bank?"

a trillion is a nice number—Lasswitz has the professor use the old definition of a trillion: a million million. This definition is still sometimes used, but nowadays the usual definition of a trillion is one thousand million: a one followed by nine zeros. In the context of the number of volumes in the Universal Library it really doesn't matter which version of a trillion you care to use.

You all know that light travels—the speed of light had been carefully measured in the decades before this story was written. At the time of its publication, a few scientists were speculating that the speed of light might be a limiting velocity in dynamics (see next chapter).

the farthest visible galaxies—Lasswitz uses the term Nebelflecken, which is the German word for 'nebulae'. I've translated it as 'galaxies', which is what Lasswitz meant in spirit. At the time he was writing, however, astronomers didn't understand the nature of galaxies.

his wife softly quoted—Mrs Wallhausen gives the second verse of Goethe's poem "Limits of humanity". I'm sure there are much better translations than the one given here!

Note As this book was going to press, I became aware of a recent English translation of "The Universal Library". See Born (2017).

Commentary

For my undergraduate physics degree, many years ago, the final set of exams contained a general paper in which we students could be asked anything. *The sky is dark at night: explain.* Or: *How many molecules of Caesar's last breath do you inhale each time you take a breath?* Or: *Derive a lower limit on the proton lifetime from the fact of your own existence.* The purpose wasn't to test whether we could remember physics facts—books were allowed into the exam hall, so we could look up whatever facts our books made available. Rather, the exam tested our ability to strip down a problem to its essence and apply mathematical arguments to reach a reasonable, if not exact, conclusion. (This was in the years B.I., before the internet. Nowadays, if smartphones were allowed, students could google the answer even to seemingly random problems such as those above.) One of the questions on my final paper essentially required us to recreate the argument made by Professor Wallhausen. I aced the paper, so I've had a soft spot for the concept of a universal library ever since.

I first came across the concept in fiction in "The Library of Babel" by Jorge Luis Borges. The Borges tale was published in 1941, almost four decades after "The Universal Library", but it's far more celebrated than Lasswitz's story. (Incidentally, Borges wrote his story while he was working—unhappily—as a librarian.) Where the scientifically-trained Lasswitz took a rigorously mathematical approach to the question of a universal library, the philosophically inclined Borges took a more metaphysical approach. In "The Library of Babel", Borges imagines a universe filled with planes of interlocking hexagonal rooms, each room having four of its walls lined with bookshelves. Spiral staircases connect the planes. Each book on the shelves is different. For Borges, each book contains 410 pages; there are 40 lines per page, 80 characters per line, and 25 different possible typographical symbols. If you run the numbers in the same way as Lasswitz did, it's easy to calculate that Borges' Library of Babel contains about 1.95×10^{1834097} different books. This is a huge number, incomparably bigger than the number of particles in the observable universe. It is, however, vastly smaller than the number of all possible books, calculated by Lasswitz, which is $10^{2000000}$. It's difficult to comprehend the size of the numbers contained in "The Universal Library" and "The Library of Babel". Perhaps a comparison with some real-world attempts at a universal library can put them into perspective.

The original universal library was the ancient Library of Alexandria, the tragic destruction of which through fire meant manuscripts of immense cultural significance were lost to the ages. The library's index was also lost in the conflagration, so it's not known for certain how many books were housed at Alexandria—experts suggest the number of scrolls would have been between 40,000 and 400,000. If the number of scrolls were at the top end of the range then the Library of Alexandria would have housed a significant fraction of the ancient world's knowledge. (It's interesting to note that an important function of a library is to classify and organize knowledge. An ancient library wouldn't necessarily have been organized in the same way as a modern library because the ancients viewed the world in a quite different way to us. What we might classify as poetry, for example, the ancients might have classified as natural science.) Moving forward in time, the present British Library contains some of the world's most significant books and manuscripts, items that are priceless. In addition to the quality of its collections, the British Library stands out in terms of quantity: it has more books than any other library except the American Library of Congress. The LOC, the world's largest library, has 32 million books and many more millions of photographs, maps, and manuscripts. Of course, even the British Library and the Library of Congress (see Fig. 7.1) now have a rival: the internet. The internet can be thought of as a library containing not just text, but images, sounds, videos, and simulations. (Indeed, one website even provides a simulation of Borge's Library of Babel— visit https://libraryofbabel.info for a disorientating glimpse of what the Library contains; see Fig. 7.2. You'll struggle to find anything of interest in it, however. As both Lasswitz and Borges emphasised, there's a problem with indexing a universal library.)

The Library of Alexandria, the British Library, the Library of Congress, and the internet. If you were to collect all the items contained in these libraries and throw in all the items from all the other libraries, public and private, that people have put together throughout history—well, the collection fill only a tiny fraction of Borge's Library of Babel. It would be vanishingly small compared to Lasswitz's Universal Library. The real world is much smaller than the world of mathematical possibility. Nevertheless, you can't deny our technological civilization is producing data at an unprecedented rate. And this opens up numerous challenges and opportunities. Let's see why.

Fig. 7.1 Reading rooms of the Library of Congress (top) and the British Library (bottom). The LOC and BL are among the largest libraries in the world, but their combined storage capacity would be insufficient to house the printed output of the Web—let alone the unimaginable vastness of the Universal Library (Credit: top—Carol M. Highsmith; bottom—Diliff)

Suppose we wanted to print out all the text that appears on the Web. How many books would we need? It's impossible to give a definitive answer, of course, but we can make an informed guess.

In 2014, scientists estimated that the internet housed 1 billion websites; the number of sites fluctuates because new websites are created and old websites are retired, but a round billion is a reasonable figure. The number of websites by itself doesn't help us, because each site can contain multiple pages, but it's possible to account for that. In 2016, scientists estimated that there

Fig. 7.2 Jonathan Basile, creator of the online universal library, standing in front of a page of text from the library (Credit: Alan Levine)

were 4.66 billion web pages. (These estimates ignored material on the so-called "Deep Web"—a corner of the internet which, for many different reasons, both legitimate and illegitimate, is not indexed by search engines. By definition, it's difficult to calculate the size of the Deep Web but experts estimate that it contains orders of magnitude more material than appears in traditionally indexed pages. For simplicity, though, let's agree to omit the Deep Web from our calculations.) In the spirit of Professor Wallhausen, let's suppose the internet contains 5 billion web pages and each web page, if printed out on paper, corresponds to 10 book pages. (I have no idea how realistic this estimate is, but it doesn't seem too unreasonable.) In this case, if you were to make a hard copy of the Web you'd end up with 50 billion printed pages. If we assume an average book has a page length of 500 then we know how big a library would have to be in order to house the "Surface" Web: the library would have to hold 100 million books. Neither the British Library nor the Library of Congress would suffice.

But our online world consists of more than just static webpages. With their tweets, Twitter users generate the equivalent of about 25,000 500-page books each day; Facebook users share 78 million links each day; around the world, about 200 billion emails are sent each day. It's as if the general public is regularly filling the Library of Congress with content. And of course digital content isn't restricted to textual material of the sort that interested Professor Wallhausen: there are graphics, maps, photos, songs, videos, simulations …

all sorts of information is now in digital form. Text-based data constitutes only a small fraction of what is stored on the internet. It's worth reiterating that any numbers we use to capture the size of the internet can't compare with the ungraspably large numbers discussed by Lasswitz in "The Universal Library". But by most real-world standards we are surely justified in saying the internet is big, and getting bigger with each passing year. In order to better quantify this, though, and understand why this trend carries with it both challenges and opportunities, we need to look at how computer scientists measure storage requirements for data.

*　*　*

Ultimately, computers work with binary digits—bits: 0 or 1, on or off, north or south. When discussing data, however, a more useful unit is the byte, which is eight binary digits long. Most computers use a byte to represent a single character—letter, number, typographical symbol. In the early days of personal computing, when people worried about how much memory was inside their machine and about the file size of their documents, units such as kilobyte and megabyte entered the common parlance. Note that the terms have two different but equally valid definitions, so there is some confusion here. A kilobyte can be 1000 bytes, as the name implies; but in computing it's convenient to work in powers of 2 so a kilobyte is often $2^{10} = 1024$ bytes. A megabyte can be 1,000,000 bytes; but it can also be $2^{20} = 1,048,576$ bytes. The difference between the two definitions increases as the size increases, but for the purposes of this discussion we needn't be concerned. We are interested in orders of magnitude, not in precise numbers.) Anyway, as many aspects of computer technology advanced along the exponential curve known as Moore's Law—with a doubling every 18 months or so—people began talking about the gigabyte (a billion bytes) and then the terabyte (a trillion bytes). My current computer contains a terabyte hard disk, a luxury unthinkable back when home computers typically came with a 360-kilobyte floppy disk drive. The pattern whereby each named unit of data is one thousand (or 1024) times greater than the previous unit continues: after the terabyte we have the petabyte, then the exabyte, zettabyte, and yottabyte. I've even seen mention of the brontobyte and geobyte. Phew.

For someone who remembers having to transfer data from computer to computer on 5.25-inch floppy disks, a unit such as the zettabyte seems ludicrously inappropriate for any realistic computing task. And yet last year, as I write, internet traffic exceeded a zettabyte; this year there's been even more traffic; next year it will be greater still. Individuals, businesses, universities, research teams … it seems as if the world is becoming a factory for generating data. As mentioned above, handling such a flood of data presents challenges—conceptual, technical, and ethical. But if we can tame the deluge then the opportunities are immense.

Let me give just one example of the challenges and opportunities of so-called Big Data. The example happens to come from astronomy, but I could have taken an example from other areas of science—or from healthcare, retail, technology … indeed, from most aspects of human endeavour.

*   *   *

We are entering a golden age of observational astronomy and cosmology. Consider, for example, the Large Synoptic Survey Telescope (LSST). When this wonderful telescope commences operations in 2022 it will consist essentially of three very large mirrors, behind which a 3.2-gigapixel digital camera will take 15-second exposures of the sky every 20 seconds. Scientists expect the camera to take 1.28 petabytes of data every year. Human astronomers simply won't be able to process that amount of data: there aren't enough eyes and brains for the task. And the LSST is just one of many observatories—operating not just throughout the electromagnetic spectrum, but also using gravitational waves and particle detectors. Some have already seen "first light", some are soon to come online. Each of them will generate such vast quantities of data that human astronomers would drown if they tried to process it manually. But if data scientists could store and index the observational data in an efficient way then *machines* would be able to mine the data for us—and make discoveries much more quickly than humans would be able to do. Indeed, machines might make serendipitous discoveries that humans themselves would miss.

This approach is already bearing fruit.

As I began to write this commentary, astronomers published a paper explaining how they trained an AI (the same sort of algorithm that Google DeepMind used to beat the world Go champion, as discussed in the previous chapter) to search for gravitational lenses. A gravitational lens can be seen when light rays from a distant galaxy are bent by the gravitational influence of an intervening galaxy; instead of observing a small disk we instead see the light of the distant galaxy smeared into arcs. The AI was trained to recognise known gravitational lenses and then asked to find lenses in a much larger data set containing millions of astronomical images. The AI quickly discovered 56 new gravitational lenses. At present, there remains a large element of human intervention in this work. Eventually, though, there'll be no need for visual inspection by humans. The discovery process will speed up immensely.

The same approach will, I'm sure, be taken in all those other areas of endeavour I mentioned above, all those other areas in which Big Data is being generated. In other words, artificial intelligence will be applied *everywhere*.

Perhaps *this* will be the future of artificial intelligence: not androids walking around with us but AIs analysing data to help us make scientific discoveries, guide political decisions, improve human health. Asimov's robot stories are typically remembered as being about androids—machines in human form. But he also wrote stories about a supercomputer called Multivac. The all-powerful Multivac was essentially a machine that had learned to navigate a useful corner of the Universal Library and excelled at Big Data problems. It acted as humanity's guide. Perhaps humans and machines will go forward together—with humans asking the questions and machines providing the answers?

I have to end this chapter with mention of Asimov's personal favourite of his own stories: "The Last Question". In the story, a technician asks Multivac a question involving the basic laws of physics: can the universal increase in entropy be reversed? Multivac ponders, then replies: "Insufficient data for meaningful answer". But Multivac doesn't forget the question, and considers it through the aeons. I won't spoil the story for you, except to say that Multivac does eventually present an answer.

*   *   *

The notion explored in "The Last Question"—whether it's possible, even for a super-advanced AI, to circumvent the laws of physics—leads us nicely into Chapter 8. The next story asks: is it possible to travel faster than light?

Notes and Further Reading

Caesar's last breath—Norman Thompson (1987) collected 137 problems asked of Bristol final-year undergraduate students over a 25-year period The questions in the book reflect a particular local tradition—short, unstructured problems that often require subtle mathematical analysis—so the book won't be for all tastes. But the questions do force you to think.

Incidentally, the question regarding Caesar's last breath provided the title to a fascinating book about Earth's atmosphere; see Kean (2017).

the "Library of Babel" by Jorge Luis Borges—Lasswitz's (1904) tale clearly influenced Borge's own story. Borges read German literature in the original, and in his essay "The total library", which was published a year before "Library of Babel", Borges explicitly refers to Lasswitz as being the first exponent of the concept. Borges story can be found in the collection *Labyrinths* (Borges 1962); his essay can be found in the collection *The Total Library* (Borges 2000).

difficult to comprehend the size of the numbers—Although the numbers appearing in this story tale seem huge beyond comprehension, mathematicians routinely work with quantities that dwarf the $10^{2,000,000}$ which Lasswitz gives as the number of possible books. In combinatorics, for example, mathematicians often use the uparrow, or $\uparrow$, to denote exponentiation. Thus $2\uparrow2 = 2^2 = 4$; $3\uparrow4 = 3^4 = 81$; and so on. A pair of arrows, $\uparrow\uparrow$, represents a tower of exponents. Thus $3\uparrow\uparrow3$ represents 3 to the power of 3 to the power of 3, or 3^{27}, which is 7,625,597,484,987. The number $3\uparrow\uparrow4$ is thus $3^{7,625,597,484,987}$, which already dwarfs the number Lasswitz calculated. For the sorts of problems in which the uparrow notation is used, even $3\uparrow\uparrow4$ would be considered insignificantly small. A discussion of the uparrow notation can be found in Webb (2018).

the ancients might have classified as natural science—For more information on ancient libraries, see Nicholls (n.d.).

the internet housed 1 billion websites—Netcraft, a UK-based internet services company, provides research data and analysis on many aspects of the internet. It has been surveying the internet since August 1995. The first time the survey exceeded one billion websites was in September 2014 (Netcraft 2014). The site internetlivestats.com (n.d.) contains a live estimate of the number of websites; at the time of writing, the number is approaching two billion.

4.66 billion web pages—This estimate of the number of web pages was made by Van den Bosch et al. (2016).

regularly filling the Library of Congress with content—The program SCIgen, created by scientists at MIT, adds a small but enjoyable trickle to this flood of textual information. An editor such as Max Burkel would dread SCIgen: the program generates nonsense that nevertheless possesses a level of superficial plausibility. The program's authors developed it to autogenerate submissions to conferences that one suspects have low submission standards. The program works. An early paper (SCIgen 2005)—with an abstract consisting of the sentences "Many physicists would agree that, had it not been for congestion control, the evaluation of web browsers might never have occurred. In fact, few hackers worldwide would disagree with the essential unification of voice-over-IP and public/private key pair. In order to solve this riddle, we confirm that SMPs can be made stochastic, cacheable, and interposable."—was accepted by a conference. Since then, SCIgen output has fooled many others. Visit https://pdos.csail.mit.edu/archive/scigen/ to generate your own computer science paper, complete with fake graphs!

astronomers published a paper—See Petrillo et al. (2017).

Asimov's personal favourite—Several Multivac stories, including "The Last Question", appear in *Robot Dreams* (Asimov 1986).

Bibliography

Asimov, I.: Robot Dreams. Berkley, New York (1986)

Borges, J.L.: Lsabyrinths (transl. D.A. Yates, J.E. Irby). New Directions, New York (1962)

Borges, J.L.: The Total Library: Non-fiction 1922–1986. Allen Lane, London (2000)

Born, E. (transl.): The Universal Library by Kurd Lasswitz. Mithila Rev.: J. Int. Sci. Fict. Fantasy. Issue 9. mithilareview.com/lasswitz_09_17 (2017)

Fadiman, C.: Fantasia Mathematica. Simon and Schuster, New York (1958)

internetlivestats.com: Total Number of Websites. http://www.internetlivestats.com/total-number-of-websites/ (n.d.)

Kean, S.: Caesar's Last Breath: Decoding the Secrets of the Air Around Us. Little, Brown & Co., New York (2017)

Lasswitz, K.: Die Universalbibliothek. Ostdeutsche Allgemeine Zeitung. **Zeitung (Dec 18), (1904)**

Netcraft: September 2014 Web Server Survey. https://news.netcraft.com/archives/2014/09/24/september-2014-web-server-survey.html (2014)

Nicholls, M.: Ancient Libraries. https://www.bl.uk/greek-manuscripts/articles/ancient-libraries (n.d.)

Petrillo, C.E., et al.: Finding strong gravitational lenses in the Kilo Degree Survey with convolutional networks. Month. Not. Roy. Astro. Soc. **472**(1), 1129–1150 (2017)

SCIgen: Rooter: A Methodology for the Typical Unification of Access Points and Redundancy. https://pdos.csail.mit.edu/archive/scigen/#about (2005)

Thompson, N.: Thinking Like a Physicist: Physics Problems for Undergraduates. Institute of Physics Publishing, Bristol (1987)

Van den Bosch, A., Bogers, T., de Kunder, M.: Estimating search engine index size variability: a 9-year longitudinal study. Scientometrics. **106**(2), (2016). https://doi.org/10.1007/s11192-016-1863-z

Webb, S.: Clash of Symbols. Springer, Berlin (2018)

8

Faster than Light Travel

The Tachypomp: A Mathematical Demonstration (E.P. Mitchell)

There was nothing mysterious about Professor Surd's dislike for me. I was the only poor mathematician in an exceptionally mathematical class. The old gentleman sought the lecture-room every morning with eagerness, and left it reluctantly. For was it not a thing of joy to find seventy young men who, individually and collectively, preferred x to XX; who had rather differentiate than dissipate; and for whom the limbs of the heavenly bodies had more attractions than those of earthly stars upon the spectacular stage?

So affairs went on swimmingly between the Professor of Mathematics and the Junior Class at Polyp University. In every man of the seventy the sage saw the logarithm of a possible LaPlace, of a Sturm, or of a Newton. It was a delightful task for him to lead them through the pleasant valleys of conic sections, and beside the still waters of the integral calculus. Figuratively speaking, his problem was not a hard one. He had only to manipulate, and eliminate, and to raise to a higher power, and the triumphant result of examination day was assured.

But I was a disturbing element, a perplexing unknown quantity, which had somehow crept into the work, and which seriously threatened to impair the accuracy of his calculations. It was a touching sight to behold the venerable mathematician as he pleaded with me not so utterly to disregard precedent in

© Springer Nature Switzerland AG 2019
S. Webb, *New Light Through Old Windows: Exploring Contemporary Science Through 12 Classic Science Fiction Tales*, Science and Fiction,
https://doi.org/10.1007/978-3-030-03195-4_8

the use of cotangents; or as he urged, with eyes almost tearful, that ordinates were dangerous things to trifle with. All in vain. More theorems went on to my cuff than into my head. Never did chalk do so much work to so little purpose. And, therefore, it came that Furnace Second was reduced to zero in Professor Surd's estimation. He looked upon me with all the horror which an unalgebraic nature could inspire. I have seen the Professor walk around an entire square rather than meet the man who had no mathematics in his soul.

For Furnace Second were no invitations to Professor Surd's house. Seventy of the class supped in delegations around the periphery of the Professor's tea-table. The seventy-first knew nothing of the charms of that perfect ellipse, with its twin bunches of fuchsias and geraniums in gorgeous precision at the two foci.

This, unfortunately enough, was no trifling deprivation. Not that I longed especially for segments of Mrs. Surd's justly celebrated lemon pies; not that the spheroidal damsons of her excellent preserving had any marked allure-ments; not even that I yearned to hear the Professor's jocose table-talk about binomials, and chatty illustrations of abstruse paradoxes. The explanation is far different. Professor Surd had a daughter. Twenty years before, he made a proposition of marriage to the present Mrs. S. He added a little Corollary to his proposition not long after. The Corollary was a girl.

Abscissa Surd was as perfectly symmetrical as Giotto's circle, and as pure, withal, as the mathematics her father taught. It was just when spring was com-ing to extract the roots of frozen-up vegetation that I fell in love with the Corollary. That she herself was not indifferent I soon had reason to regard as a self-evident truth.

The sagacious reader will already recognize nearly all the elements necessary to a well-ordered plot. We have introduced a heroine, inferred a hero, and constructed a hostile parent after the most approved model. A movement for the story, a Deus ex machina, is alone lacking. With considerable satisfaction I can promise a perfect novelty in this line, a Deus ex machina never before offered to the public.

It would be discounting ordinary intelligence to say that I sought with unwearying assiduity to figure my way into the stern father's good-will; that never did dullard apply himself to mathematics more patiently than I; that never did faithfulness achieve such meagre reward. Then I engaged a private tutor. His instructions met with no better success.

My tutor's name was Jean Marie Rivarol. He was a unique Alsatian—though Gallic in name, thoroughly Teuton in nature; by birth a Frenchman, by education a German. His age was thirty; his profession, omniscience; the wolf at his door, poverty; the skeleton in his closet, a consuming but unrequited passion. The most recondite principles of practical science were his toys; the deepest intricacies of abstract science his diversions. Problems which were foreordained mysteries to me were to him as clear as Tahoe water. Perhaps this very fact will explain our lack of success in the relation of tutor and pupil; perhaps the failure is alone due to my own unmitigated stupidity. Rivarol had hung about the skirts of the University for several years; supplying his few wants by writing for scientific journals, or by giving assistance to students who, like myself, were characterized by a plethora of purse and a paucity of ideas; cooking, studying and sleeping in his attic lodgings; and prosecuting queer experiments all by himself.

We were not long discovering that even this eccentric genius could not transplant brains into my deficient skull. I gave over the struggle in despair. An unhappy year dragged its slow length around. A gloomy year it was, brightened only by occasional interviews with Abscissa, the Abbie of my thoughts and dreams.

Commencement day was coming on apace. I was soon to go forth, with the rest of my class, to astonish and delight a waiting world. The Professor seemed to avoid me more than ever. Nothing but the conventionalities, I think kept him from shaping his treatment of me on the basis of unconcealed disgust.

At last, in the very recklessness of despair, I resolved to see him, plead with him, threaten him if need be, and risk all my fortunes on one desperate chance. I wrote him a somewhat defiant letter, stating my aspirations, and, as I flattered myself, shrewdly giving him a week to get over the first shock of horrified surprise. Then I was to call and learn my fate.

During the week of suspense I nearly worried myself into a fever. It was first crazy hope, and then saner despair. On Friday evening, when I presented myself at the Professor's door, I was such a haggard, sleepy, dragged-out spectre, that even Miss Jocasta, the harsh-favored maiden sister of the Surd's, admitted me with commiserate regard, and suggested pennyroyal tea.

Professor Surd was at a faculty meeting. Would I wait? Yes, till all was blue, if need be. Miss Abbie?

Abscissa had gone to Wheelborough to visit a school-friend. The aged maiden hoped I would make myself comfortable, and departed to the unknown haunts which knew Jocasta's daily walk.

Comfortable! But I settled myself in a great uneasy chair and waited, with the contradictory spirit common to such junctures, dreading every step lest it should herald the man whom, of all men, I wished to see.

I had been there at least an hour, and was growing right drowsy.

At length Professor Surd came in. He sat down in the dusk opposite me, and I thought his eyes glinted with malignant pleasure as he said, abruptly:

"So, young man, you think you are a fit husband for my girl?"

I stammered some inanity about making up in affection what I lacked in merit; about my expectations, family and the like. He quickly interrupted me. "You misapprehend me, sir. Your nature is destitute of those mathematical perceptions and acquirements which are the only sure foundations of character. You have no mathematics in you. You are fit for treason, stratagems, and spoils.—Shakespeare. Your narrow intellect cannot understand and appreciate a generous mind. There is all the difference between you and a Surd, if I may say it, which intervenes between an infinitesimal and an infinite. Why, I will even venture to say that you do not comprehend the Problem of the Couriers!"

I admitted that the Problem of the Couriers should be classed rather without my list of accomplishments than within it. I regretted this fault very deeply, and suggested amendment. I faintly hoped that my fortune would be such—

"Money!" he impatiently exclaimed. "Do you seek to bribe a Roman Senator with a penny whistle? Why, boy, do you parade your paltry wealth, which, expressed in mills, will not cover ten decimal places, before the eyes of a man who measures the planets in their orbits, and close crowds infinity itself?"

I hastily disclaimed any intention of obtruding my foolish dollars, and he went on:

"Your letter surprised me not a little. I thought *you* would be the last person in the world to presume to an alliance here. But having a regard for you personally"—and again I saw malice twinkle in his small eyes—"and still more regard for Abscissa's happiness, I have decided that you shall have her—upon conditions. Upon conditions," he repeated, with a half-smothered sneer.

"What are they?" cried I, eagerly enough. "Only name them."

"Well, sir," he continued, and the deliberation of his speech seemed the very refinement of cruelty, "you have only to prove yourself worthy an alliance with a mathematical family. You have only to accomplish a task which I shall presently give you. Your eyes ask me what it is. I will tell you. Distinguish yourself in that noble branch of abstract science in which, you cannot but acknowledge, you are at present sadly deficient. I will place Abscissa's hand in yours whenever you shall come before me and square the circle to my satisfaction. No! That is too easy a condition. I should cheat myself. Say perpetual motion. How do you like that? Do you think it lies within the range of your mental capabilities? You don't smile. Perhaps your talents don't run in the way of perpetual motion. Several people have found that theirs didn't. I'll give you another chance. We were speaking of the Problem of the Couriers, and I think you expressed a desire to know more of that ingenious question. You shall have the opportunity. Sit down some day, when you have nothing else to do, and discover the principle of infinite speed. I mean the law of motion which shall accomplish an infinitely great distance in an infinitely short time. You may mix in a little practical mechanics, if you choose. Invent some method of taking the tardy Courier over his road at the rate of sixty miles a minute. Demonstrate me this discovery (when you have made it!) mathematically, and approximate it practically, and Abscissa is yours. Until you can, I will thank you to trouble neither myself nor her."

I could stand his mocking no longer. I stumbled mechanically out of the room, and out of the house. I even forgot my hat and gloves. For an hour I walked in the moonlight. Gradually I succeeded to a more hopeful frame of mind. This was due to my ignorance of mathematics. Had I understood the real meaning of what he asked, I should have been utterly despondent.

Perhaps this problem of sixty miles a minute was not so impossible after all. At any rate I could attempt, though I might not succeed. And Rivarol came to my mind. I would ask him. I would enlist his knowledge to accompany my own devoted perseverance. I sought his lodgings at once.

The man of science lived in the fourth story, back. I had never been in his room before. When I entered, he was in the act of filling a beer mug from a carboy labelled *Aqua fortis*.

"Seat you," he said. "No, not in that chair. That is my Petty Cash Adjuster." But he was a second too late. I had carelessly thrown myself into a chair of seductive appearance. To my utter amazement it reached out two skeleton arms and clutched me with a grasp against which I struggled in vain. Then a skull stretched itself over my shoulder and grinned with ghastly familiarity close to my face.

Rivarol came to my aid with many apologies. He touched a spring somewhere and the Petty Cash Adjuster relaxed its horrid hold. I placed myself gingerly in a plain cane-bottomed rocking-chair, which Rivarol assured me was a safe location.

"That seat," he said, "is an arrangement upon which I much felicitate myself. I made it at Heidelberg. It has saved me a vast deal of small annoyance. I consign to its embraces the friends who bore, and the visitors who exasperate, me. But it is never so useful as when terrifying some tradesman with an insignificant account. Hence the pet name which I have facetiously given it. They are invariably too glad to purchase release at the price of a bill receipted. Do you well apprehend the idea?"

While the Alsatian diluted his glass of *Aqua fortis*, shook into it an infusion of bitters, and tossed off the bumper with apparent relish, I had time to look around the strange apartment.

The four corners of the room were occupied respectively by a turning-lathe, a Rhumkorff Coil, a small steam-engine and an orrery in stately motion. Tables, shelves, chairs and floor supported an odd aggregation of tools, retorts, chemicals, gas-receivers, philosophical instruments, boots, flasks, paper-collar boxes, books diminutive and books of preposterous size. There were plaster busts of Aristotle, Archimedes, and Comte, while a great drowsy owl was blinking away, perched on the benign brow of Martin Farquhar Tupper. "He always roosts there when he proposes to slumber," explained my tutor. "You are a bird of no ordinary mind. *Schlafen Sie wohl.*"

Through a closet door, half open, I could see a human-like form covered with a sheet. Rivarol caught my glance.

"That," said he, "will be my masterpiece. It is a Microcosm, an Android, as yet only partially complete. And why not? Albertus Magnus constructed an image perfect to talk metaphysics and confute the schools. So did Sylvester II.; so did Robertus Greathead. Roger Bacon made a brazen head that held discourses. But the first named of these came to destruction. Thomas Aquinas got wrathful at some of its syllogisms and smashed its head. The idea is reasonable enough. Mental action will yet be reduced to laws as definite as those which govern the physical. Why should not I accomplish a manikin which shall preach as original discourses as the Rev. Dr. Allchin, or talk poetry as mechanically as Paul Anapest? My Android can already work problems in vulgar fractions and compose sonnets. I hope to teach it the Positive Philosophy."

Out of the bewildering confusion of his effects Rivarol produced two pipes and filled them. He handed one to me.

"And here," he said, "I live and am tolerably comfortable. When my coat wears out at the elbows I seek the tailor and am measured for another. When I am hungry I promenade myself to the butcher's and bring home a pound or so of steak, which I cook very nicely in three seconds by this oxy-hydrogen flame. Thirsty, perhaps, I send for a carboy of *Aqua fortis*. But I have it charged, all charged. My spirit is above any small pecuniary transaction. I loathe your dirty greenbacks, and never handle what they call scrip."

"But are you never pestered with bills?" I asked. "Don't the creditors worry your life out?"

"Creditors!" gasped Rivarol. "I have learned no such word in your very admirable language. He who will allow his soul to be vexed by creditors is a relic of an imperfect civilization. Of what use is science if it cannot avail a man who has accounts current? Listen. The moment you or any one else enters the outside door this little electric bell sounds me warning. Every successive step on Mrs. Grimier's staircase is a spy and informer vigilant for my benefit. The first step is trod upon. That trusty first step immediately telegraphs your weight. Nothing could be simpler. It is exactly like any platform scale. The weight is registered up here upon this dial. The second step records the size of my visitor's feet. The third his height, the fourth his complexion, and so on. By the time he reaches the top of the first flight I have a pretty accurate description of him right here at my elbow, and quite a margin of time for deliberation and action. Do you follow me? It is plain enough. Only the ABC of my science."

"I see all that," I said, "but I don't see how it helps you any. The knowledge that a creditor is coming won't pay his bill. You can't escape unless you jump out of the window."

Rivarol laughed softly. "I will tell you. You shall see what becomes of any poor devil who goes to demand money of me—of a man of science. Ha! ha! It pleases me. I was seven weeks perfecting my Dun Suppressor. Did you know"—he whispered exultingly—"did you know that there is a hole through the earth's centre? Physicists have long suspected it; I was the first to find it. You have read how Rhuyghens, the Dutch navigator, discovered in Kerguellen's Land an abysmal pit which fourteen hundred fathoms of plumb-line failed to sound. Herr Tom, that hole has no bottom! It runs from one surface of the earth to the antipodal surface. It is diametric. But where is the antipodal spot? You stand upon it. I learned this by the merest chance. I was deep-digging in Mrs. Grimler's cellar, to bury a poor cat I had sacrificed in a galvanic experiment, when the earth under my spade crumbled, caved in, and wonder-stricken I stood upon the brink of a yawning shaft. I dropped a coal-hod in. It went down, down down, bounding and rebounding. In two hours and a quarter that coal-hod came up again. I caught it and restored it to the angry Grimler. Just think a minute. The coal-hod went down, faster and faster, till it reached the centre of the earth. There it would stop, were it not for acquired momentum. Beyond the centre its journey was relatively upward, toward the opposite surface of the globe. So, losing velocity, it went slower and slower till it reached that surface. Here it came to rest for a second and then fell back again, eight thousand odd miles, into my hands. Had I not interfered with it, it would have repeated its journey, time after time, each trip of shorter extent, like the diminishing oscillations of a pendulum, till it finally came to eternal rest at the centre of the sphere. I am not slow to give a practical application to any such grand discovery. My Dun Suppressor was born of it. A trap, just outside my chamber door: a spring in here: a creditor on the trap:—need I say more?"

"But isn't it a trifle inhuman?" I mildly suggested. "Plunging an unhappy being into a perpetual journey to and from Kerguellen's Land, without a moment's warning."

"I give them a chance. When they come up the first time I wait at the mouth of the shaft with a rope in hand. If they are reasonable and will come to terms, I fling them the line. If they perish, 'tis their own fault. Only," he added, with a melancholy smile, "the centre is getting so plugged up with creditors that I am afraid there soon will be no choice whatever for 'em."

By this time I had conceived a high opinion of my tutor's ability. If anybody could send me waltzing through space at an infinite speed, Rivarol could do it. I filled my pipe and told him the story. He heard with grave and patient attention. Then, for full half an hour, he whiffed away in silence. Finally he spoke.

"The ancient cipher has overreached himself. He has given you a choice of two problems, both of which he deems insoluble. Neither of them is insoluble. The only gleam of intelligence Old Cotangent showed was when he said that squaring the circle was too easy. He was right. It would have given you your *Liebchen* in five minutes. I squared the circle before I discarded pantalets. I will show you the work—but it would be a digression, and you are in no mood for digressions. Our first chance, therefore, lies in perpetual motion. Now, my good friend, I will frankly tell you that, although I have compassed this interesting problem, I do not choose to use it in your behalf. I too, Herr Tom, have a heart. The loveliest of her sex frowns upon me. Her somewhat mature charms are not for Jean Marie Rivarol. She has cruelly said that her years demand of me filial rather than connubial regard. Is love a matter of years or of eternity? This question did I put to the cold, yet lovely Jocasta."

"Jocasta Surd!" I remarked in surprise, "Abscissa's aunt!"

"The same," he said, sadly. "I will not attempt to conceal that upon the maiden Jocasta my maiden heart has been bestowed. Give me your hand, my nephew in affliction as in affection!"

Rivarol dashed away a not discreditable tear, and resumed:

"My only hope lies in this discovery of perpetual motion. It will give me the fame, the wealth. Can Jocasta refuse these? If she can, there is only the trap-door and—Kerguellen's Land!"

I bashfully asked to see the perpetual-motion machine. My uncle in affliction shook his head.

"At another time," he said. "Suffice it at present to say, that it is something upon the principle of a woman's tongue. But you see now why we must turn in your case to the alternative condition—infinite speed. There are several ways in which this may be accomplished, theoretically. By the lever, for instance. Imagine a lever with a very long and a very short arm. Apply power to the shorter arm which will move it with great velocity. The end of the long arm will move much faster. Now keep shortening the short arm and lengthening the long one, and as you approach infinity in their difference of length, you approach infinity in the speed of the long arm. It would be difficult to demonstrate this practically to the Professor. We must seek another solution. Jean Marie will meditate. Come to me in a fortnight. Good-night. But stop! Have you the money—*das Geld*?"

"Much more than I need."

"Good! Let us strike hands. Gold and Knowledge; Science and Love. What may not such a partnership achieve? We go to conquer thee, Abscissa. *Vorwärts*!"

When, at the end of a fortnight, I sought Rivarol's chamber, I passed with some little trepidation over the terminus of the Air Line to Kerguellen's Land, and evaded the extended arms of the Petty Cash Adjuster. Rivarol drew a mug of ale for me, and filled himself a retort of his own peculiar beverage.

"Come," he said at length. "Let us drink success to the TACHYPOMP."

"The TACHYPOMP?"

"Yes. Why not? *Tachu*, quickly, and *pempo*, *pepompa* to send. May it send you quickly to your wedding-day. Abscissa is yours. It is done. When shall we start for the prairies?"

"Where is it?" I asked, looking in vain around the room for any contrivance which might seem calculated to advance matrimonial prospects.

"It is here," and he gave his forehead a significant tap. Then he held forth didactically.

"There is force enough in existence to yield us a speed of sixty miles a minute, or even more. All we need is the knowledge how to combine and apply it. The wise man will not attempt to make some great force yield some great speed. He will keep adding the little force to the little force, making each little force yield its little speed, until an aggregate of little forces shall be a great force, yielding an aggregate of little speeds, a great speed. The difficulty is not in aggregating the forces; it lies in the corresponding aggregation of the speeds. One musket-ball will go, say a mile. It is not hard to increase the force of muskets to a thousand, yet the thousand musket-balls will go no farther, and no faster, than the one. You see, then, where our trouble lies. We cannot readily add speed to speed, as we add force to force. My discovery is simply the utilization of a principle which extorts an increment of speed from each increment of power. But this is the metaphysics of physics. Let us be practical or nothing.

"When you have walked forward, on a moving train, from the rear car, toward the engine, did you ever think what you were really doing?"

"Why, yes, I have generally been going to the smoking-car to have a cigar."

"Tut, tut—not that! I mean, did it ever occur to you on such an occasion, that absolutely you were moving faster than the train? The train passes the telegraph poles at the rate of thirty miles an hour, say. You walk toward the smoking-car at the rate of four miles an hour. Then *you* pass the telegraph poles at the rate of thirty-four miles. Your absolute speed is the speed of the engine, plus the speed of your own locomotion. Do you follow me?"

I began to get an inkling of his meaning, and told him so.

"Very well. Let us advance a step. Your addition to the speed of the engine is trivial, and the space in which you can exercise it, limited. Now suppose two stations, A and B, two miles distant by the track. Imagine a train of platform cars, the last car resting at station A. The train is a mile long, say. The engine is therefore within a mile of station B. Say the train can move a mile in ten minutes. The last car, having two miles to go, would reach B in twenty minutes, but the engine, a mile ahead, would get there in ten. You jump on the last car, at A, in a prodigious hurry to reach Abscissa, who is at B. If you stay on the last car it will be twenty long minutes before you see her. But the engine reaches B and the fair lady in ten. You will be a stupid reasoner, and an indifferent lover, if you don't put for the engine over those platform cars, as fast as your legs will carry you. You can run a mile, the length of the train, in ten minutes. Therefore, you reach Abscissa when the engine does, or in ten minutes—ten minutes sooner than if you had lazily sat down upon the rear car and talked politics with the brakeman. You have diminished the time by one half. You have added your speed to that of the locomotive to some purpose. *Nicht wahr?*"

I saw it perfectly; much plainer, perhaps, for his putting in the clause about Abscissa.

He continued:

"This illustration, though a slow one, leads up to a principle which may be carried to any extent. Our first anxiety will be to spare your legs and wind. Let us suppose that the two miles of track are perfectly straight, and make our train one platform car, a mile long, with parallel rails laid upon its top. Put a little dummy engine on these rails, and let it run to and fro along the platform car, while the platform car is pulled along the ground track. Catch the idea? The dummy takes your place. But it can run its mile much faster. Fancy that our locomotive is strong enough to pull the platform car over the two miles in two minutes. The dummy can attain the same speed. When the engine reaches B in one minute, the dummy, having gone a mile a-top the platform car, reaches B also. We have so combined the speeds of those two engines as to accomplish two miles in one minute. Is this all we can do? Prepare to exercise your imagination."

I lit my pipe.

"Still two miles of straight track, between A and B. On the track a long platform car, reaching from A to within a quarter of a mile of B. We will now discard ordinary locomotives and adopt as our motive power a series of compact magnetic engines, distributed underneath the platform car, all along its length."

"I don't understand those magnetic engines."

"Well, each of them consists of a great iron horseshoe, rendered alternately a magnet and not a magnet by an intermittent current of electricity from a battery, this current in its turn regulated by clock-work. When the horseshoe is in the circuit, it is a magnet, and it pulls its clapper toward it with enormous power. When it is out of the circuit, the next second, it is not a magnet, and it lets the clapper go. The clapper, oscillating to and fro, imparts a rotatory motion to a fly-wheel, which transmits it to the drivers on the rails. Such are our motors. They are no novelty, for trial has proved them practicable.

"With a magnetic engine for every truck of wheels, we can reasonably expect to move our immense car, and to drive it along at a speed, say, of a mile a minute.

"The forward end, having but a quarter of a mile to go, will reach B in fifteen seconds. We will call this platform car number 1. On top of number 1 are laid rails on which another platform car, number 2, a quarter of a mile shorter than number 1, is moved in precisely the same way. Number 2, in its turn, is surmounted by number 3, moving independently of the tiers beneath, and a quarter of a mile shorter than number 2. Number 2 is a mile and a half long; number 3 a mile and a quarter. Above, on successive levels, are number 4, a mile long; number 5, three quarters of a mile; number 6, half a mile; number 7, a quarter of a mile, and number 8, a short passenger car, on top of all.

"Each car moves upon the car beneath it, independently of all the others, at the rate of a mile a minute. Each car has its own magnetic engines. Well, the train being drawn up with the latter end of each car resting against a lofty bumping-post at A, Tom Furnace, the gentlemanly conductor, and Jean Marie Rivarol, engineer, mount by a long ladder to the exalted number 8. The complicated mechanism is set in motion. What happens?

"Number 8 runs a quarter of a mile in fifteen seconds and reaches the end of number 7. Meanwhile number 7 has run a quarter of a mile in the same time and reached the end of number 6; number 6, a quarter of a mile in fifteen seconds, and reached the end of number 5; number 5, the end of number 4; number 4, of number 3; number 3, of number 2; number 2, of number 1. And number 1, in fifteen seconds, has gone its quarter of a mile along the ground track, and has reached station B. All this has been done in fifteen seconds. Wherefore, numbers 1, 2, 3, 4, 5, 6, 7, and 8 come to rest against the bumping-post at B, at precisely the same second. We, in number 8, reach B just when number 1 reaches it. In other words, we accomplish two miles in fifteen seconds. Each of the eight cars, moving at the rate of a mile a minute, has contributed a quarter of a mile to our journey, and has done its work in fifteen seconds. All the eight did their work at once, during the same fifteen seconds. Consequently we have been whizzed through the air at the somewhat startling speed of seven and a half seconds to the mile. This is the Tachypomp. Does it justify the name?"

Although a little bewildered by the complexity of cars, I apprehended the general principle of the machine. I made a diagram, and understood it much better. "You have merely improved on the idea of my moving faster than the train when I was going to the smoking car?"

"Precisely. So far we have kept within the bounds of the practicable. To satisfy the Professor, you can theorize in something after this fashion: If we double the number of cars, thus decreasing by one half the distance which each has to go, we shall attain twice the speed. Each of the sixteen cars will have but one eighth of a mile to go. At the uniform rate we have adopted, the two miles can be done in seven and a half instead of fifteen seconds. With thirty-two cars, and a sixteenth of a mile, or twenty rods difference in their length, we arrive at the speed of a mile in less than two seconds; with sixty-four cars, each travelling but ten rods, a mile under the second. More than sixty miles a minute! If this isn't rapid enough for the Professor, tell him to go on, increasing the number of his cars and diminishing the distance each one has to run. If sixty-four cars yield a speed of a mile inside the second, let him fancy a Tachypomp of six hundred and forty cars, and amuse himself calculating the rate of car number 640. Just whisper to him that when he has an infinite number of cars with an infinitesimal difference in their lengths, he will have obtained that infinite speed for which he seems to yearn. Then demand Abscissa."

I wrung my friend's hand in silent and grateful admiration. I could say nothing.

"You have listened to the man of theory," he said proudly. "You shall now behold the practical engineer. We will go to the west of the Mississippi and find some suitably level locality. We will erect thereon a model Tachypomp. We will summon thereunto the professor, his daughter, and why not his fair sister Jocasta, as well? We will take them a journey which shall much astonish the venerable Surd. He shall place Abscissa's digits in yours and bless you both with an algebraic formula. Jocasta shall contemplate with wonder the genius of Rivarol. But we have much to do. We must ship to St. Joseph the vast amount of material to be employed in the construction of the Tachypomp. We must engage a small army of workmen to effect that construction, for we are to annihilate time and space. Perhaps you had better see your bankers."

I rushed impetuously to the door. There should be no delay.

"Stop! stop! *Um Gottes Willen*, stop!" shrieked Rivarol. "I launched my butcher this morning and I haven't bolted the——"

But it was too late. I was upon the trap. It swung open with a crash, and I was plunged down, down, down! I felt as if I were falling through illimitable space. I remember wondering, as I rushed through the darkness, whether I should reach Kerguellen's Land or stop at the centre. It seemed an eternity. Then my course was suddenly and painfully arrested.

I opened my eyes. Around me were the walls of Professor Surd's study. Under me was a hard, unyielding plane which I knew too well was Professor Surd's study floor. Behind me was the black, slippery, hair-cloth chair which had belched me forth, much as the whale served Jonah. In front of me stood Professor Surd himself, looking down with a not unpleasant smile.

"Good-evening, Mr. Furnace. Let me help you up. You look tired, sir. No wonder you fell asleep when I kept you so long waiting. Shall I get you a glass of wine? No? By the way, since receiving your letter I find that you are a son of my old friend, Judge Furnace. I have made inquiries, and see no reason why you should not make Abscissa a good husband."

Still I can see no reason why the Tachypomp should not have succeeded. Can you?

Edward Page Mitchell (1852–1927) was editor for *The Sun*, a high-circulation daily newspaper in New York City; he also wrote SF stories for the paper. In his stories Mitchell developed many of the major themes of science fiction: he wrote of invisibility and time travel before Wells; he wrote the earliest known tale of teleportation; he wrote about a cyborg eight decades before the term itself was invented. Unfortunately, his stories for *The Sun* appeared anonymously. This might explain the relative obscurity of his name.

Commentary

Mitchell's playful story contains a number of impossibilities. But the impossibilities are not all of the same order.

Take Mitchell's explanation of the Dun Suppressor. (A dun, incidentally, is an archaic term for a debt collector.) The construction of such a device is entirely impossible for present-day engineers. Nothing in the laws of physics, however, rules out the fabrication of a tube linking two antipodal points. If humankind possessed greater technical capacity (incredibly strong, heat-resistant materials for the tunnel walls, for example); plus a means of alleviating the Coriolis effect, a by-product of Earth's rotation that would cause the traveller to be slammed into the tunnel walls; plus the political desire and economic wherewithal to drill down to the centre of the Earth and then straight out again … well, I suppose a Dun Suppressor could be made. And yes, after being pushed into the tube, the poor victim would indeed oscillate between opposite points on the Earth's surface like a mass oscillates on the end of a spring. So the Dun Suppressor is a *practical* impossibility but it's not a *fundamental* impossibility.

In the story, however, Professor Surd considers setting the unwelcome suitor one of three challenges—each of which we now know to be *truly* impossible.

The construction of a perpetual motion machine—that's one challenge Surd ponders whether to set. As far back as the 15th century, Leonardo da Vinci was scathing about the possibility of perpetual motion. He wrote: "Oh ye seekers after perpetual motion, how many vain chimeras have you pursued? Go and take your place with the alchemists." Modern scientists understand that perpetual motion in an isolated system violates either the first law of thermodynamics, the second law of thermodynamics, or both. The first law is essentially the conservation of energy: you can't create energy from nothing. The second law is essentially the observation that heat flows spontaneously from a hot place to a colder place. (In the previous chapter I mentioned how Asimov's Multivac computer wrestled with the question of whether the second law of thermodynamics could be reversed.) Thus the laws of physics rule out the construction of a perpetual motion machine. Building such a device is a *fundamental* impossibility. Mitchell might not have kept abreast of contemporary developments in thermodynamics, but he probably knew that a perpetual motion machine was impossible.

Another challenge Surd considers—squaring the circle—is often used as a synonym for impossibility. Ancient geometers first posed the question: if you possess only a compass and a straight edge can you, in a finite number of steps, construct a square that has the same area as a given circle? In 1882, a German mathematician proved the answer is: no. So this is a *mathematical* impossibility—and although the proof did not appear until eight years after publication of his story, Mitchell probably suspected the challenge could not be met.

And then there's the third of Surd's challenges: show mathematically how to propel an object to an infinite speed in a finite time. For a practical demonstration Surd merely wants to see a method for travelling at 60 miles per minute. Such a speed must have seemed effectively infinite for Mitchell, who was writing at a time when steam trains provided the fastest method of travel: a stately 1 mile per minute, flat out. Nowadays, of course, space rockets can go much faster than Mitchell would have dreamed. So, since it's eminently possible to accelerate a vehicle to a speed of 60 miles per minute, let's stick with the *spirit* of Surd's challenge. Is it possible to get an object to move at infinite speed?

Mitchell seems to consider this challenge to be along the same lines as the construction of a Dun Suppressor: difficult to build in practice but possible in principle. However, as Albert Einstein showed 31 years after the publication of Mitchell's story, the laws of physics make the task impossible. The suitor's tachypomp could not possibly work. Here's why.

* * *

Imagine a physics laboratory floating freely out in space, far away from the gravitational pull of planets, the particle wind from stars, and the magnetic fields from the galaxy. In this ideal situation physicists inside the lab don't have to worry about any complicating effects from outside the lab when they measure something. Suppose physicists inside such a lab decide to make a measurement of some fundamental physical quantity (the speed of light, say). Here's an interesting fact about how our universe is put together: the result of a measurement made *now* will be the same as the result of a measurement made *tomorrow*. It doesn't matter *when* they make the measurement—in these ideal circumstances the outcome is always the same. In other words, the laws of physics don't depend upon time. Similarly, if they make a measurement *here* and repeat the measurement *there* then they'll obtain the same result. It doesn't matter *where* they make the measurement. In other words, the laws of physics don't depend upon location. And if they make a measurement with the lab pointing in one direction then they'll get the same result with the lab pointing some other direction. It doesn't matter *how* they are oriented when they make the measurement. In other words, the laws of physics don't depend upon direction.

In 1905, Einstein extended this idea regarding the universality of the laws of physics. As long as you are moving smoothly, at a constant velocity, it doesn't matter how *fast* you are moving when you make a measurement of some fundamental physical quantity. You'll measure the same laws of physics no matter what your speed happens to be.

The theory Einstein built around this idea goes by the name of relativity, even though the theory at its heart is about an invariance: the invariance of the laws of physics with respect to velocity. The reason for the term "relativity" is the following. If you're inside a space-based laboratory that's moving smoothly then you can't perform an experiment inside the lab to tell whether you are moving. You can consider yourself to be at rest. If you look out the lab windows and see a similar lab whizz smoothly past then you can say the other object is moving *relative to you*; scientists in the other lab can say *you* are moving *relative to them*. Nevertheless, physicists inside labs that are moving smoothly relative to one another will measure the same value of fundamental quantities. In particular: *all observers measure the same value for the speed of light.*

From Einstein's realisation—namely, that all smoothly moving observers measure the same value for the speed of light—it follows that observers who move at different velocities must make different measurements of space and time (because, by definition, to measure the velocity of something you must measure the space traversed in a given time). In our everyday world people agree on *when* an event happens and they agree on the *distance* between events—but that's only because we only ever move slowly. When velocities are small in comparison with the speed of light (and remember that the fastest aircraft moves at only a tiny fraction of light speed) the effects predicted by Einstein's relativity are unobservably small. But when velocities are a significant fraction of light speed, observers disagree about the passage of time and they disagree about distances in space. This conclusion might seem deeply counterintuitive—we tend to think in terms of time and space being absolute; they are supposed to be quantities we can all agree upon—but it happens to be how the universe is put together. Since 1905, countless experiments have confirmed Einstein's insights.

What has all this got to do with the impossibility of a tachypomp? Well, the tachypomp works by adding velocities—and another counterintuitive aspect of relativity is the way in which velocities add.

Fig. 8.1 A sketch of the tachypomp, which appeared alongside an early re-publication of Mitchell's story. In a tower of vehicles, each moving relative to the one below, will the topmost vehicle be able to exceed the speed of light? Mitchell's character Rivarol said yes; Einstein told us no (Credit: Reginald Birch)

Suppose you are in one of those smoothly moving space-based laboratories, far away from any complicating effects, and you see another lab move smoothly past. In other words, the two labs are moving relative to one another, and occupants of both labs are entitled to believe they are at rest and it's the *other* lab that's moving. So far, so good. Now, suppose you see the other lab move past you at an even pace of 5 km/s, and you observer that an experimenter in the other lab fires a ball out of a front-facing cannon at a speed of 10 km/s. Unless you have exceptionally accurate equipment you, in your lab, will measure the speed of the ball to be 5 + 10 = 15 km/s. That's how velocities combine in our everyday world: they simply add.

Mitchell, in his story, writes that if a train travels past telegraph poles at 30 mph, and a passenger walks along the train corridor at 4 mph, then the passenger moves past the telegraph poles at 30 mph + 4 mph = 34 mph. This is the basis of the tachypomp (see Fig. 8.1), and it assumes that the rule for adding small velocities holds when large velocities are involved. *But the rule doesn't hold!*

Returning to our space lab, suppose the other lab zooms past you at $0.5c$. (Here, c is the standard symbol for the speed of light. So we're supposing that the lab moves past you at half the speed of light.) And instead of firing a ball, the experimenter in the other lab uses a torch to fire a beam of light out in front. According to the normal rule for adding velocities, you should see the light beam moving at $0.5c + c = 1.5c$. But that's clearly wrong. It must be wrong because everyone always agrees about the speed of light: you see the light beam travel at c not at $1.5c$. Light travels at light speed, irrespective of how fast the original source of the light happens to be moving. Clearly, then, there must be a different rule for adding velocities when large speeds are involved. This isn't a textbook, so I shan't go through the formula for adding relativistic velocities; you can easily look it up if you are interested. It's enough to give some examples.

If a spacecraft moves past you travelling at $0.5c$ and it fires a projectile forward at $0.5c$ then you as a stationary observer see the projectile moving at $0.8c$; if the spacecraft is moving at $0.75c$ when it fires a projectile at $0.5c$ then you observe the projectile travel at a speed $0.909c$; if the spacecraft moves at $0.8c$ and the projectile is fired forward at $0.8c$ then you observe the projectile to be moving at $0.976c$. The formula is such that when you add together *any* two sub-c velocities the result is always less than c.

Had Mitchell been able to apply the correct formula for the relativistic addition of velocities to his tachypomp he would have discovered that all observers conclude that the device *doesn't* reach an infinite speed in a finite time. Indeed, it doesn't—and it never can—even reach light speed.

Einstein's picture of a relativistic universe is contrary to our common, everyday experience. Einstein told us that clocks run slow when they move; that rods contract along the direction of travel; that, no matter how much you try, you can't accelerate an object to faster-than-light speed. We don't observe these effects in our low-speed world, but relativistic effects always come into play whenever velocities are large. We can't avoid those effects. They are an inevitable consequence of the way the universe is put together. A tachypomp *can't* move faster than the speed of light. It's an impossibility.

*   *   *

You can no more build a tachypomp than you can a perpetual motion machine: the laws of physics rule them both out. Chapter 9 examines yet another phenomenon that runs foul of universal laws: antigravity.

Notes and Further Reading

Leonardo da Vinci was scathing about the possibility of perpetual motion— Leonardo's quote can be seen in his notebooks; see Leonardo (1906) for an English translation, which now can be accessed online.

Fig. 8.2 A photograph of the Michelson–Morley interferometer, taken around about the time of the famous experiment. The experiment took place at what is now Case Western Reserve University. The interferometer sent two beams of light in perpendicular directions; mirrors reflected the light back, and the two beams recombined. Michelson and Morley expected to see a changing interference pattern as Earth moved through the aether. Instead, they observed a static pattern—one of the most famous null results in all of science (Credit: Case Western Reserve University)

the laws of physics rule out the construction of a perpetual motion machine— Atkins (2010) gives one of the best introductions to thermodynamics, but there are countless other volumes on the subject. Once you understand the laws of thermodynamics, you can immediately see why a perpetual motion machine is impossible.

a German mathematician proved the answer is: no—Long before 1882, mathematicians had shown that if π were a transcendental number (in other words, that it is not a root of a polynomial with rational coefficients) then it would be impossible to square the circle. In 1882, Carl Louis Ferdinand von Lindemann showed that π is indeed a transcendental number—and so, by implication, that squaring the circle is impossible. For a non-technical introduction to transcendental numbers see, for example, Higgins (2008).

countless experiments have confirmed Einstein's insights—The first experiment to hint at these ideas took place as early as 1887. The Michelson–Morley experiment, see Fig. 8.2, compared the speed of light in perpendicular directions. At the time of the experiment, some 13 years after Mitchell's (1874) story, scientists continued to believe in a quantity known as the aether—a medium through which light could propagate. If the aether did exist then scientists would be able to measure a different speed of light, depending on whether a light beam moved "with" or "against" the aether. (If the aether did exist then a tachypomp would make sense so, given that Mitchell was writing thirteen years before the Michelson–Morley experiment, we should cut him some slack!) The Michelson–Morley experiment found no difference in speed, a null result suggesting the aether does not exist and the speed of light is constant. It's not clear how much this experiment featured in Einstein's development of special relativity, but it undoubtedly helped scientists accept the theory. After the publication of Einstein's theory in 1905, all experiments have been consistent with the ideas of special relativity.

the standard symbol for the speed of light—Webb (2018) describes the story behind using c as the symbol for light speed.

the formula for adding relativistic velocities—You can find innumerable books about special relativity; some take an historical approach, some a mathematical approach, yet others a diagrammatic approach. Take your pick. All of them will explain the correct formula for adding velocities. Personally, I was taught the theory via French (1968). A nice modern approach, for those wanting to teach themselves the subject, is Susskind and Friedman (2017).

Bibliography

Atkins, P.: The Laws of Thermodynamics: A Very Short Introduction. Oxford University Press, Oxford (2010)

French, A.P.: Special Relativity. Norton, New York (1968)

Higgins, P.M.: Number Story: From Counting to Cryptography. Copernicus, New York (2008)

Leonardo: Leonardo Da Vinci's Note-books: Arranged and Rendered into English, with Introduction by Edward McCurdy. Duckworth, London (1906)

Mitchell, E.P.: The tachypomp: a mathematical demonstration. The Sun. January 1874

Susskind, L., Friedman, A.: Special Relativity and Classical Field Theory: The Theoretical Minimum. Basic, New York (2017)

Webb, S.: Clash of Symbols. Springer, Berlin (2018)

9

Antigravity

A Tale of Negative Gravity (Frank R. Stockton)

My wife and I were staying at a small town in northern Italy; and on a certain pleasant afternoon in spring we had taken a walk of six or seven miles to see the sun set behind some low mountains to the west of the town. Most of our walk had been along a hard, smooth highway, and then we turned into a series of narrower roads, sometimes bordered by walls, and sometimes by light fences of reed or cane. Nearing the mountain, to a low spur of which we intended to ascend, we easily scaled a wall about four feet high, and found ourselves upon pasture-land, which led, sometimes by gradual ascents, and sometimes by bits of rough climbing, to the spot we wished to reach. We were afraid we were a little late, and therefore hurried on, running up the grassy hills, and bounding briskly over the rough and rocky places. I carried a knapsack strapped firmly to my shoulders, and under my wife's arm was a large, soft basket of a kind much used by tourists. Her arm was passed through the handles and around the bottom of the basket, which she pressed closely to her side. This was the way she always carried it. The basket contained two bottles of wine, one sweet for my wife, and another a little acid for myself. Sweet wines give me a headache.

When we reached the grassy bluff, well known thereabouts to lovers of sunset views, I stepped immediately to the edge to gaze upon the scene, but my wife sat down to take a sip of wine, for she was very thirsty; and then,

© Springer Nature Switzerland AG 2019
S. Webb, *New Light Through Old Windows: Exploring Contemporary Science Through 12 Classic Science Fiction Tales*, Science and Fiction,
https://doi.org/10.1007/978-3-030-03195-4_9

leaving her basket, she came to my side. The scene was indeed one of great beauty. Beneath us stretched a wide valley of many shades of green, with a little river running through it, and red-tiled houses here and there. Beyond rose a range of mountains, pink, pale green, and purple where their tips caught the reflection of the setting sun, and of a rich gray-green in shadows. Beyond all was the blue Italian sky, illumined by an especially fine sunset.

My wife and I are Americans, and at the time of this story were middle-aged people and very fond of seeing in each other's company whatever there was of interest or beauty around us. We had a son about twenty-two years old, of whom we were also very fond; but he was not with us, being at that time a student in Germany. Although we had good health, we were not very robust people, and, under ordinary circumstances, not much given to long country tramps. I was of medium size, without much muscular development, while my wife was quite stout, and growing stouter.

The reader may, perhaps, be somewhat surprised that a middle-aged couple, not very strong, or very good walkers, the lady loaded with a basket containing two bottles of wine and a metal drinking-cup, and the gentleman carrying a heavy knapsack, filled with all sorts of odds and ends, strapped to his shoulders, should set off on a seven-mile walk, jump over a wall, run up a hillside, and yet feel in very good trim to enjoy a sunset view. This peculiar state of things I will proceed to explain.

I had been a professional man, but some years before had retired upon a very comfortable income. I had always been very fond of scientific pursuits, and now made these the occupation and pleasure of much of my leisure time. Our home was in a small town; and in a corner of my grounds I built a laboratory, where I carried on my work and my experiments. I had long been anxious to discover the means not only of producing, but of retaining and controlling, a natural force, really the same as centrifugal force, but which I called negative gravity. This name I adopted because it indicated better than any other the action of the force in question, as I produced it. Positive gravity attracts everything toward the centre of the earth. Negative gravity, therefore, would be that power which repels everything from the centre of the earth, just as the negative pole of a magnet repels the needle, while the positive pole attracts it. My object was, in fact, to store centrifugal force and to render it constant, controllable, and available for use. The advantages of such a discovery could scarcely be described. In a word, it would lighten the burdens of the world.

I will not touch upon the labors and disappointments of several years. It is enough to say that at last I discovered a method of producing, storing, and controlling negative gravity.

The mechanism of my invention was rather complicated, but the method of operating it was very simple. A strong metallic case, about eight inches long, and half as wide, contained the machinery for producing the force; and this was put into action by means of the pressure of a screw worked from the outside. As soon as this pressure was produced, negative gravity began to be evolved and stored, and the greater the pressure the greater the force. As the screw was moved outward, and the pressure diminished, the force decreased, and when the screw was withdrawn to its fullest extent, the action of negative gravity entirely ceased. Thus this force could be produced or dissipated at will to such degrees as might be desired, and its action, so long as the requisite pressure was maintained, was constant.

When this little apparatus worked to my satisfaction I called my wife into my laboratory and explained to her my invention and its value. She had known that I had been at work with an important object, but I had never told her what it was. I had said that if I succeeded I would tell her all, but if I failed she need not be troubled with the matter at all. Being a very sensible woman, this satisfied her perfectly. Now I explained everything to her—the construction of the machine, and the wonderful uses to which this invention could be applied. I told her that it could diminish, or entirely dissipate, the weight of objects of any kind. A heavily loaded wagon, with two of these instruments fastened to its sides, and each screwed to a proper force, would be so lifted and supported that it would press upon the ground as lightly as an empty cart, and a small horse could draw it with ease. A bale of cotton, with one of these machines attached, could be handled and carried by a boy. A car, with a number of these machines, could be made to rise in the air like a balloon. Everything, in fact, that was heavy could be made light; and as a great part of labor, all over the world, is caused by the attraction of gravitation, so this repellent force, wherever applied, would make weight less and work easier. I told her of many, many ways in which the invention might be used, and would have told her of many more if she had not suddenly burst into tears.

"The world has gained something wonderful," she exclaimed, between her sobs, "but I have lost a husband!"

"What do you mean by that?" I asked, in surprise.

"I haven't minded it so far," she said, "because it gave you something to do, and it pleased you, and it never interfered with our home pleasures and our home life. But now that is all over. You will never be your own master again. It will succeed, I am sure, and you may make a great deal of money, but we don't need money. What we need is the happiness which we have always had until now. Now there will be companies, and patents, and lawsuits, and experiments, and people calling you a humbug, and other people saying they discovered it long ago, and all sorts of persons coming to see you, and you'll be obliged to go to all sorts of places, and you will be an altered man, and we shall never be happy again. Millions of money will not repay us for the happiness we have lost."

These words of my wife struck me with much force. Before I had called her my mind had begun to be filled and perplexed with ideas of what I ought to do now that the great invention was perfected. Until now the matter had not troubled me at all. Sometimes I had gone backward and sometimes forward, but, on the whole, I had always felt encouraged. I had taken great pleasure in the work, but I had never allowed myself to be too much absorbed by it. But now everything was different. I began to feel that it was due to myself and to my fellow-beings that I should properly put this invention before the world. And how should I set about it? What steps should I take? I must make no mistakes. When the matter should become known hundreds of scientific people might set themselves to work; how could I tell but that they might discover other methods of producing the same effect? I must guard myself against a great many things. I must get patents in all parts of the world. Already, as I have said, my mind began to be troubled and perplexed with these things. A turmoil of this sort did not suit my age or disposition. I could not but agree with my wife that the joys of a quiet and contented life were now about to be broken into.

"My dear," said I, "I believe, with you, that the thing will do us more harm than good. If it were not for depriving the world of the invention I would throw the whole thing to the winds. And yet," I added, regretfully, "I had expected a great deal of personal gratification from the use of this invention."

"Now listen," said my wife, eagerly; "don't you think it would be best to do this: use the thing as much as you please for your own amusement and satisfaction, but let the world wait? It has waited a long time, and let it wait a little longer. When we are dead let Herbert have the invention. He will then be old enough to judge for himself whether it will be better to take advantage of it for his own profit, or simply to give it to the public for nothing. It would be cheating him if we were to do the latter, but it would also be doing him a great wrong if we were, at his age, to load him with such a heavy responsibility. Besides, if he took it up, you could not help going into it, too."

I took my wife's advice. I wrote a careful and complete account of the invention, and, sealing it up, I gave it to my lawyers to be handed to my son after my death. If he died first, I would make other arrangements. Then I determined to get all the good and fun out of the thing that was possible without telling any one anything about it. Even Herbert, who was away from home, was not to be told of the invention.

The first thing I did was to buy a strong leathern knapsack, and inside of this I fastened my little machine, with a screw so arranged that it could be worked from the outside. Strapping this firmly to my shoulders, my wife gently turned the screw at the back until the upward tendency of the knapsack began to lift and sustain me. When I felt myself so gently supported and upheld that I seemed to weigh about thirty or forty pounds, I would set out for a walk. The knapsack did not raise me from the ground, but it gave me a very buoyant step. It was no labor at all to walk; it was a delight, an ecstasy. With the strength of a man and the weight of a child, I gayly strode along. The first day I walked half a dozen miles at a very brisk pace, and came back without feeling in the least degree tired. These walks now became one of the greatest joys of my life. When nobody was looking, I would bound over a fence, sometimes just touching it with one hand, and sometimes not touching it at all. I delighted in rough places. I sprang over streams. I jumped and I ran. I felt like Mercury himself.

I now set about making another machine, so that my wife could accompany me in my walks; but when it was finished she positively refused to use it. "I can't wear a knapsack," she said, "and there is no other good way of fastening it to me. Besides, everybody about here knows I am no walker, and it would only set them talking."

I occasionally made use of this second machine, but I will give only one instance of its application. Some repairs were needed to the foundation-walls of my barn, and a two-horse wagon, loaded with building-stone, had been brought into my yard and left there. In the evening, when the men had gone away, I took my two machines and fastened them, with strong chains, one on each side of the loaded wagon. Then, gradually turning the screws, the wagon was so lifted that its weight became very greatly diminished. We had an old donkey which used to belong to Herbert, and which was now occasionally used with a small cart to bring packages from the station. I went into the barn and put the harness on the little fellow, and, bringing him out to the wagon, I attached him to it. In this position he looked very funny with a long pole sticking out in front of him and the great wagon behind him. When all was ready I touched him up; and, to my great delight, he moved off with the two-horse load of stone as easily as if he were drawing his own cart. I led him out into the public road, along which he proceeded without difficulty. He was an opinionated little beast, and sometimes stopped, not liking the peculiar manner in which he was harnessed; but a touch of the switch made him move on, and I soon turned him and brought the wagon back into the yard. This determined the success of my invention in one of its most important uses, and with a satisfied heart I put the donkey into the stable and went into the house.

Our trip to Europe was made a few months after this, and was mainly on our son Herbert's account. He, poor fellow, was in great trouble, and so, therefore, were we. He had become engaged, with our full consent, to a young lady in our town, the daughter of a gentleman whom we esteemed very highly. Herbert was young to be engaged to be married, but as we felt that he would never find a girl to make him so good a wife, we were entirely satisfied, especially as it was agreed on all hands that the marriage was not to take place for some time. It seemed to us that, in marrying Janet Gilbert, Herbert would secure for himself, in the very beginning of his career, the most important element of a happy life. But suddenly, without any reason that seemed to us justifiable, Mr. Gilbert, the only surviving parent of Janet, broke off the match; and he and his daughter soon after left the town for a trip to the West.

This blow nearly broke poor Herbert's heart. He gave up his professional studies and came home to us, and for a time we thought he would be seriously ill. Then we took him to Europe, and after a Continental tour of a month or two we left him, at his own request, in Göttingen, where he thought it would do him good to go to work again. Then we went down to the little town in Italy where my story first finds us. My wife had suffered much in mind and body on her son's account, and for this reason I was anxious that she should take outdoor exercise, and enjoy as much as possible the bracing air of the country. I had brought with me both my little machines. One was still in my knapsack, and the other I had fastened to the inside of an enormous family trunk. As one is obliged to pay for nearly every pound of his baggage on the Continent, this saved me a great deal of money. Everything heavy was packed into this great trunk—books, papers, the bronze, iron, and marble relics we had picked up, and all the articles that usually weigh down a tourist's baggage. I screwed up the negative-gravity apparatus until the trunk could be handled with great ease by an ordinary porter. I could have made it weigh nothing at all, but this, of course, I did not wish to do. The lightness of my baggage, however, had occasioned some comment, and I had overheard remarks which were not altogether complimentary about people travelling around with empty trunks; but this only amused me.

Desirous that my wife should have the advantage of negative gravity while taking our walks, I had removed the machine from the trunk and fastened it inside of the basket, which she could carry under her arm. This assisted her wonderfully. When one arm was tired she put the basket under the other, and thus, with one hand on my arm, she could easily keep up with the free and buoyant steps my knapsack enabled me to take. She did not object to long tramps here, because nobody knew that she was not a walker, and she always carried some wine or other refreshment in the basket, not only because it was pleasant to have it with us, but because it seemed ridiculous to go about carrying an empty basket.

There were English-speaking people stopping at the hotel where we were, but they seemed more fond of driving than walking, and none of them offered to accompany us on our rambles, for which we were very glad. There was one man there, however, who was a great walker. He was an Englishman, a member of an Alpine Club, and generally went about dressed in a knickerbocker suit, with gray woollen stockings covering an enormous pair of calves. One evening this gentleman was talking to me and some others about the ascent of the Matterhorn, and I took occasion to deliver in pretty strong language my opinion upon such exploits. I declared them to be useless, foolhardy, and, if the climber had any one who loved him, wicked.

"Even if the weather should permit a view," I said, "what is that compared to the terrible risk to life? Under certain circumstances," I added (thinking of a kind of waistcoat I had some idea of making, which, set about with little negative-gravity machines, all connected with a conveniently handled screw, would enable the wearer at times to dispense with his weight altogether), "such ascents might be divested of danger, and be quite admissible; but ordinarily they should be frowned upon by the intelligent public."

The Alpine Club man looked at me, especially regarding my somewhat slight figure and thinnish legs.

"It's all very well for you to talk that way," he said, "because it is easy to see that you are not up to that sort of thing."

"In conversations of this kind," I replied, "I never make personal allusions; but since you have chosen to do so, I feel inclined to invite you to walk with me to-morrow to the top of the mountain to the north of this town."

"I'll do it," he said, "at any time you choose to name." And as I left the room soon afterward I heard him laugh.

The next afternoon, about two o'clock, the Alpine Club man and myself set out for the mountain.

"What have you got in your knapsack?" he said.

"A hammer to use if I come across geological specimens, a field-glass, a flask of wine, and some other things."

"I wouldn't carry any weight, if I were you," he said.

"Oh, I don't mind it," I answered, and off we started.

The mountain to which we were bound was about two miles from the town. Its nearest side was steep, and in places almost precipitous, but it sloped away more gradually toward the north, and up that side a road led by devious windings to a village near the summit. It was not a very high mountain, but it would do for an afternoon's climb.

"I suppose you want to go up by the road," said my companion.

"Oh no," I answered, "we won't go so far around as that. There is a path up this side, along which I have seen men driving their goats. I prefer to take that."

"All right, if you say so," he answered, with a smile; "but you'll find it pretty tough."

After a time he remarked:

"I wouldn't walk so fast, if I were you."

"Oh, I like to step along briskly," I said. And briskly on we went.

My wife had screwed up the machine in the knapsack more than usual, and walking seemed scarcely any effort at all. I carried a long alpenstock, and when we reached the mountain and began the ascent, I found that with the help of this and my knapsack I could go uphill at a wonderful rate. My companion had taken the lead, so as to show me how to climb. Making a détour over some rocks, I quickly passed him and went ahead. After that it was impossible for him to keep up with me. I ran up steep places, I cut off the windings of the path by lightly clambering over rocks, and even when I followed the beaten track my step was as rapid as if I had been walking on level ground.

"Look here!" shouted the Alpine Club man from below, "you'll kill yourself if you go at that rate! That's no way to climb mountains."

"It's my way!" I cried. And on I skipped.

Twenty minutes after I arrived at the summit my companion joined me, puffing, and wiping his red face with his handkerchief.

"Confound it!" he cried, "I never came up a mountain so fast in my life."

"You need not have hurried," I said, coolly.

"I was afraid something would happen to you," he growled, "and I wanted to stop you. I never saw a person climb in such an utterly absurd way."

"I don't see why you should call it absurd," I said, smiling with an air of superiority. "I arrived here in a perfectly comfortable condition, neither heated nor wearied."

He made no answer, but walked off to a little distance, fanning himself with his hat and growling words which I did not catch. After a time I proposed to descend.

"You must be careful as you go down," he said. "It is much more dangerous to go down steep places than to climb up."

"I am always prudent," I answered, and started in advance. I found the descent of the mountain much more pleasant than the ascent. It was positively exhilarating. I jumped from rocks and bluffs eight and ten feet in height, and touched the ground as gently as if I had stepped down but two feet. I ran down steep paths, and, with the aid of my alpenstock, stopped myself in an instant. I was careful to avoid dangerous places, but the runs and jumps I made were such as no man had ever made before upon that mountain-side. Once only I heard my companion's voice.

"You'll break your —— neck!" he yelled.

"Never fear!" I called back, and soon left him far above.

When I reached the bottom I would have waited for him, but my activity had warmed me up, and as a cool evening breeze was beginning to blow I thought it better not to stop and take cold. Half an hour after my arrival at the hotel I came down to the court, cool, fresh, and dressed for dinner, and just in time to meet the Alpine man as he entered, hot, dusty, and growling.

"Excuse me for not waiting for you," I said; but without stopping to hear my reason, he muttered something about waiting in a place where no one would care to stay, and passed into the house.

There was no doubt that what I had done gratified my pique and tickled my vanity.

"I think now," I said, when I related the matter to my wife, "that he will scarcely say that I am not up to that sort of thing."

"I am not sure," she answered, "that it was exactly fair. He did not know how you were assisted."

"It was fair enough," I said. "He is enabled to climb well by the inherited vigor of his constitution and by his training. He did not tell me what methods of exercise he used to get those great muscles upon his legs. I am enabled to climb by the exercise of my intellect. My method is my business and his method is his business. It is all perfectly fair."

Still she persisted:

"He thought that you climbed with your legs, and not with your head." And now, after this long digression, necessary to explain how a middle-aged couple of slight pedestrian ability, and loaded with a heavy knapsack and basket, should have started out on a rough walk and climb, fourteen miles in all, we will return to ourselves, standing on the little bluff and gazing out upon the sunset view. When the sky began to fade a little we turned from it and prepared to go back to the town.

"Where is the basket?" I said.

"I left it right here," answered my wife. "I unscrewed the machine and it lay perfectly flat."

"Did you afterward take out the bottles?" I asked, seeing them lying on the grass.

"Yes, I believe I did. I had to take out yours in order to get at mine."

"Then," said I, after looking all about the grassy patch on which we stood, "I am afraid you did not entirely unscrew the instrument, and that when the weight of the bottles was removed the basket gently rose into the air."

"It may be so," she said, lugubriously. "The basket was behind me as I drank my wine."

"I believe that is just what has happened," I said. "Look up there! I vow that is our basket!"

I pulled out my field-glass and directed it at a little speck high above our heads. It was the basket floating high in the air. I gave the glass to my wife to look, but she did not want to use it.

"What shall I do?" she cried. "I can't walk home without that basket. It's perfectly dreadful!" And she looked as if she was going to cry.

"Do not distress yourself," I said, although I was a good deal disturbed myself. "We shall get home very well. You shall put your hand on my shoulder, while I put my arm around you. Then you can screw up my machine a good deal higher, and it will support us both. In this way I am sure that we shall get on very well."

We carried out this plan, and managed to walk on with moderate comfort. To be sure, with the knapsack pulling me upward, and the weight of my wife pulling me down, the straps hurt me somewhat, which they had not done before. We did not spring lightly over the wall into the road, but, still clinging to each other, we clambered awkwardly over it. The road for the most part declined gently toward the town, and with moderate ease we made our way along it. But we walked much more slowly than we had done before, and it was quite dark when we reached our hotel. If it had not been for the light inside the court it would have been difficult for us to find it. A travelling-carriage was standing before the entrance, and against the light. It was necessary to pass around it, and my wife went first. I attempted to follow her, but, strange to say, there was nothing under my feet. I stepped vigorously, but only wagged my legs in the air. To my horror I found that I was rising in the air! I soon saw, by the light below me, that I was some fifteen feet from the ground. The carriage drove away, and in the darkness I was not noticed. Of course I knew what had happened. The instrument in my knapsack had been screwed up to such an intensity, in order to support both myself and my wife, that when her weight was removed the force of the negative gravity was sufficient to raise me from the ground. But I was glad to find that when I had risen to the height I have mentioned I did not go up any higher, but hung in the air, about on a level with the second tier of windows of the hotel.

I now began to try to reach the screw in my knapsack in order to reduce the force of the negative gravity; but, do what I would, I could not get my hand to it. The machine in the knapsack had been placed so as to support me in a well-balanced and comfortable way; and in doing this it had been impossible to set the screw so that I could reach it. But in a temporary arrangement of the kind this had not been considered necessary, as my wife always turned the screw for me until sufficient lifting power had been attained. I had intended, as I have said before, to construct a negative-gravity waistcoat, in which the screw should be in front, and entirely under the wearer's control; but this was a thing of the future.

When I found that I could not turn the screw I began to be much alarmed. Here I was, dangling in the air, without any means of reaching the ground. I could not expect my wife to return to look for me, as she would naturally suppose I had stopped to speak to some one. I thought of loosening myself from the knapsack, but this would not do, for I should fall heavily, and either kill myself or break some of my bones. I did not dare to call for assistance, for if any of the simple-minded inhabitants of the town had discovered me floating in the air they would have taken me for a demon, and would probably have shot at me. A moderate breeze was blowing, and it wafted me gently down the street. If it had blown me against a tree I would have seized it, and have endeavored, so to speak, to climb down it; but there were no trees. There was a dim street-lamp here and there, but reflectors above them threw their light upon the pavement, and none up to me. On many accounts I was glad that the night was so dark, for, much as I desired to get down, I wanted no one to see me in my strange position, which, to any one but myself and wife, would be utterly unaccountable. If I could rise as high as the roofs I might get on one of them, and, tearing off an armful of tiles, so load myself that I would be heavy enough to descend. But I did not rise to the eaves of any of the houses. If there had been a telegraph-pole, or anything of the kind that I could have clung to, I would have taken off the knapsack, and would have endeavored to scramble down as well as I could. But there was nothing I could cling to. Even the water-spouts, if I could have reached the face of the houses, were embedded in the walls. At an open window, near which I was slowly blown, I saw two little boys going to bed by the light of a dim candle. I was dreadfully afraid that they would see me and raise an alarm. I actually came so near to the window that I threw out one foot and pushed against the wall with such force that I went nearly across the street. I thought I caught sight of a frightened look on the face of one of the boys; but of this I am not sure, and I heard no cries. I still floated, dangling, down the street. What was to be done? Should I call out? In that case, if I were not shot or stoned, my strange predicament, and the secret of my invention, would be exposed to the world. If I did not do this, I must either let myself drop and be killed or mangled, or hang there and die. When, during the course of the night, the air became more rarefied, I might rise higher and higher, perhaps to an altitude of one or two hundred feet. It would then be impossible for the people to reach me and get me down, even if they were convinced that I was not a demon. I should then expire, and when the birds of the air had eaten all of me that they could devour, I should forever hang above the unlucky town, a dangling skeleton with a knapsack on its back.

Such thoughts were not reassuring, and I determined that if I could find no means of getting down without assistance, I would call out and run all risks; but so long as I could endure the tension of the straps I would hold out, and hope for a tree or a pole. Perhaps it might rain, and my wet clothes would then become so heavy that I would descend as low as the top of a lamp-post. As this thought was passing through my mind I saw a spark of light upon the street approaching me. I rightly imagined that it came from a tobacco-pipe, and presently I heard a voice. It was that of the Alpine Club man. Of all people in the world I did not want him to discover me, and I hung as motionless as possible. The man was speaking to another person who was walking with him.

"He is crazy beyond a doubt," said the Alpine man. "Nobody but a maniac could have gone up and down that mountain as he did! He hasn't any muscles, and one need only look at him to know that he couldn't do any climbing in a natural way. It is only the excitement of insanity that gives him strength."

The two now stopped almost under me, and the speaker continued: "Such things are very common with maniacs. At times they acquire an unnatural strength which is perfectly wonderful. I have seen a little fellow struggle and fight so that four strong men could not hold him."

Then the other person spoke.

"I am afraid what you say is too true," he remarked. "Indeed, I have known it for some time."

At these words my breath almost stopped. It was the voice of Mr. Gilbert, my townsman, and the father of Janet. It must have been he who had arrived in the travelling-carriage. He was acquainted with the Alpine Club man, and they were talking of me. Proper or improper, I listened with all my ears.

"It is a very sad case," Mr. Gilbert continued. "My daughter was engaged to marry his son, but I broke off the match. I could not have her marry the son of a lunatic, and there could be no doubt of his condition. He has been seen—a man of his age, and the head of a family—to load himself up with a heavy knapsack, which there was no earthly necessity for him to carry, and go skipping along the road for miles, vaulting over fences and jumping over rocks and ditches like a young calf or a colt. I myself saw a most heartrending instance of how a kindly man's nature can be changed by the derangement of his intellect. I was at some distance from his house, but I plainly saw him harness a little donkey which he owns to a large two-horse wagon loaded with stone, and beat and lash the poor little beast until it drew the heavy load some distance along the public road. I would have remonstrated with him on this horrible cruelty, but he had the wagon back in his yard before I could reach him."

"Oh, there can be no doubt of his insanity," said the Alpine Club man, "and he oughtn't to be allowed to travel about in this way. Some day he will pitch his wife over a precipice just for the fun of seeing her shoot through the air."

"I am sorry he is here," said Mr. Gilbert, "for it would be very painful to meet him. My daughter and I will retire very soon, and go away as early tomorrow morning as possible, so as to avoid seeing him."

And then they walked back to the hotel.

For a few moments I hung, utterly forgetful of my condition, and absorbed in the consideration of these revelations. One idea now filled my mind. Everything must be explained to Mr. Gilbert, even if it should be necessary to have him called to me, and for me to speak to him from the upper air.

Just then I saw something white approaching me along the road. My eyes had become accustomed to the darkness, and I perceived that it was an upturned face. I recognized the hurried gait, the form; it was my wife. As she came near me, I called her name, and in the same breath entreated her not to scream. It must have been an effort for her to restrain herself, but she did it.

"You must help me to get down," I said, "without anybody seeing us."

"What shall I do?" she whispered.

"Try to catch hold of this string."

Taking a piece of twine from my pocket, I lowered one end to her. But it was too short; she could not reach it. I then tied my handkerchief to it, but still it was not long enough.

"I can get more string, or handkerchiefs," she whispered, hurriedly.

"No," I said; "you could not get them up to me. But, leaning against the hotel wall, on this side, in the corner, just inside of the garden gate, are some fishing-poles. I have seen them there every day. You can easily find them in the dark. Go, please, and bring me one of those."

The hotel was not far away, and in a few minutes my wife returned with a fishing-pole. She stood on tiptoe, and reached it high in air; but all she could do was to strike my feet and legs with it. My most frantic exertions did not enable me to get my hands low enough to touch it.

"Wait a minute," she said; and the rod was withdrawn.

I knew what she was doing. There was a hook and line attached to the pole, and with womanly dexterity she was fastening the hook to the extreme end of the rod. Soon she reached up, and gently struck at my legs. After a few attempts the hook caught in my trousers, a little below my right knee. Then there was a slight pull, a long scratch down my leg, and the hook was stopped by the top of my boot. Then came a steady downward pull, and I felt myself descending. Gently and firmly the rod was drawn down; carefully the lower end was kept free from the ground; and in a few moments my ankle was seized with a vigorous grasp. Then some one seemed to climb up me, my feet touched the ground, an arm was thrown around my neck, the hand of another arm was busy at the back of my knapsack, and I soon stood firmly in the road, entirely divested of negative gravity.

"Oh that I should have forgotten," sobbed my wife, "and that I should have dropped your arms and let you go up into the air! At first I thought that you had stopped below, and it was only a little while ago that the truth flashed upon me. Then I rushed out and began looking up for you. I knew that you had wax matches in your pocket, and hoped that you would keep on striking them, so that you would be seen."

"But I did not wish to be seen," I said, as we hurried to the hotel; "and I can never be sufficiently thankful that it was you who found me and brought me down. Do you know that it is Mr. Gilbert and his daughter who have just arrived? I must see him instantly. I will explain it all to you when I come upstairs."

I took off my knapsack and gave it to my wife, who carried it to our room, while I went to look for Mr. Gilbert. Fortunately I found him just as he was about to go up to his chamber. He took my offered hand, but looked at me sadly and gravely.

"Mr. Gilbert," I said, "I must speak to you in private. Let us step into this room. There is no one here."

"My friend," said Mr. Gilbert, "it will be much better to avoid discussing this subject. It is very painful to both of us, and no good can come from talking of it."

"You cannot now comprehend what it is I want to say to you," I replied. "Come in here, and in a few minutes you will be very glad that you listened to me."

My manner was so earnest and impressive that Mr. Gilbert was constrained to follow me, and we went into a small room called the smoking-room, but in which people seldom smoked, and closed the door. I immediately began my statement. I told my old friend that I had discovered, by means that I need not explain at present, that he had considered me crazy, and that now the most important object of my life was to set myself right in his eyes. I thereupon gave him the whole history of my invention, and explained the reason of the actions that had appeared to him those of a lunatic. I said nothing about the little incident of that evening. That was a mere accident, and I did not care now to speak of it.

Mr. Gilbert listened to me very attentively.

"Your wife is here?" he asked, when I had finished.

"Yes," I said; "and she will corroborate my story in every item, and no one could ever suspect her of being crazy. I will go and bring her to you."

In a few minutes my wife was in the room, had shaken hands with Mr. Gilbert, and had been told of my suspected madness. She turned pale, but smiled.

"He did act like a crazy man," she said, "but I never supposed that anybody would think him one." And tears came into her eyes.

"And now, my dear," said I, "perhaps you will tell Mr. Gilbert how I did all this."

And then she told him the story that I had told.

Mr. Gilbert looked from the one to the other of us with a troubled air. "Of course I do not doubt either of you, or rather I do not doubt that you believe what you say. All would be right if I could bring myself to credit that such a force as that you speak of can possibly exist."

"That is a matter," said I, "which I can easily prove to you by actual demonstration. If you can wait a short time, until my wife and I have had something to eat—for I am nearly famished, and I am sure she must be—I will set your mind at rest upon that point."

"I will wait here," said Mr. Gilbert, "and smoke a cigar. Don't hurry yourselves. I shall be glad to have some time to think about what you have told me."

When we had finished the dinner, which had been set aside for us, I went upstairs and got my knapsack, and we both joined Mr. Gilbert in the smoking-room. I showed him the little machine, and explained, very briefly, the principle of its construction. I did not give any practical demonstration of its action, because there were people walking about the corridor who might at any moment come into the room; but, looking out of the window, I saw that the night was much clearer. The wind had dissipated the clouds, and the stars were shining brightly.

"If you will come up the street with me," said I to Mr. Gilbert, "I will show you how this thing works."

"That is just what I want to see," he answered.

"I will go with you," said my wife, throwing a shawl over her head. And we started up the street.

When we were outside the little town I found the starlight was quite sufficient for my purpose. The white roadway, the low walls, and objects about us, could easily be distinguished.

"Now," said I to Mr. Gilbert, "I want to put this knapsack on you, and let you see how it feels, and how it will help you to walk." To this he assented with some eagerness, and I strapped it firmly on him. "I will now turn this screw," said I, "until you shall become lighter and lighter."

"Be very careful not to turn it too much," said my wife, earnestly.

"Oh, you may depend on me for that," said I, turning the screw very gradually.

Mr. Gilbert was a stout man, and I was obliged to give the screw a good many turns.

"There seems to be considerable hoist in it," he said, directly. And then I put my arms around him, and found that I could raise him from the ground.

"Are you lifting me?" he exclaimed, in surprise. "Yes; I did it with ease," I answered. "Upon—my—word!" ejaculated Mr. Gilbert.

I then gave the screw a half-turn more, and told him to walk and run. He started off, at first slowly, then he made long strides, then he began to run, and then to skip and jump. It had been many years since Mr. Gilbert had skipped and jumped. No one was in sight, and he was free to gambol as much as he pleased. "Could you give it another turn?" said he, bounding up to me. "I want to try that wall." I put on a little more negative gravity, and he vaulted over a five-foot wall with great ease. In an instant he had leaped back into the road, and in two bounds was at my side. "I came down as light as a cat," he said. "There was never anything like it." And away he went up the road, taking steps at least eight feet long, leaving my wife and me laughing heartily at the preternatural agility of our stout friend. In a few minutes he was with us again. "Take it off," he said. "If I wear it any longer I shall want one myself, and then I shall be taken for a crazy man, and perhaps clapped into an asylum."

"Now," said I, as I turned back the screw before unstrapping the knapsack, "do you understand how I took long walks, and leaped and jumped; how I ran uphill and downhill, and how the little donkey drew the loaded wagon?"

"I understand it all," cried he. "I take back all I ever said or thought about you, my friend."

"And Herbert may marry Janet?" cried my wife.

"May marry her!" cried Mr. Gilbert. "Indeed, he shall marry her, if I have anything to say about it! My poor girl has been drooping ever since I told her it could not be."

My wife rushed at him, but whether she embraced him or only shook his hands I cannot say; for I had the knapsack in one hand and was rubbing my eyes with the other.

"But, my dear fellow," said Mr. Gilbert, directly, "if you still consider it to your interest to keep your invention a secret, I wish you had never made it. No one having a machine like that can help using it, and it is often quite as bad to be considered a maniac as to be one."

"My friend," I cried, with some excitement, "I have made up my mind on this subject. The little machine in this knapsack, which is the only one I now possess, has been a great pleasure to me. But I now know it has also been of the greatest injury indirectly to me and mine, not to mention some direct inconvenience and danger, which I will speak of another time. The secret lies with us three, and we will keep it. But the invention itself is too full of temptation and danger for any of us."

As I said this I held the knapsack with one hand while I quickly turned the screw with the other. In a few moments it was high above my head, while I with difficulty held it down by the straps. "Look!" I cried. And then I released my hold, and the knapsack shot into the air and disappeared into the upper gloom.

I was about to make a remark, but had no chance, for my wife threw herself upon my bosom, sobbing with joy.

"Oh, I am so glad—so glad!" she said. "And you will never make another?"

"Never another!" I answered.

"And now let us hurry in and see Janet," said my wife.

"You don't know how heavy and clumsy I feel," said Mr. Gilbert, striving to keep up with us as we walked back. "If I had worn that thing much longer, I should never have been willing to take it off!"

Janet had retired, but my wife went up to her room.

"I think she has felt it as much as our boy," she said, when she rejoined me. "But I tell you, my dear, I left a very happy girl in that little bedchamber over the garden."

And there were three very happy elderly people talking together until quite late that evening. "I shall write to Herbert to-night," I said, when we separated, "and tell him to meet us all in Geneva. It will do the young man no harm if we interrupt his studies just now."

"You must let me add a postscript to the letter," said Mr. Gilbert, "and I am sure it will require no knapsack with a screw in the back to bring him quickly to us."

And it did not.

There is a wonderful pleasure in tripping over the earth like a winged Mercury, and in feeling one's self relieved of much of that attraction of gravitation which drags us down to earth and gradually makes the movement of our bodies but weariness and labor. But this pleasure is not to be compared, I think, to that given by the buoyancy and lightness of two young and loving hearts, reunited after a separation which they had supposed would last forever.

What became of the basket and the knapsack, or whether they ever met in upper air, I do not know. If they but float away and stay away from ken of mortal man, I shall be satisfied.

And whether or not the world will ever know more of the power of negative gravity depends entirely upon the disposition of my son Herbert, when—after a good many years, I hope—he shall open the packet my lawyers have in keeping.

[*Note.*—It would be quite useless for any one to interview my wife on this subject, for she has entirely forgotten how my machine was made. And as for Mr. Gilbert, he never knew.]

Frank Richard Stockton (1834–1902) was an American writer, popular towards the end of the 19th century for his humorous fairy tales and stories for children. His most famous tale was *The Lady, or the Tiger?* (1882), about a man who is punished for having an affair with the king's daughter. The man is put before two doors and told to open one. Behind one door is a beautiful lady-in-waiting; if he finds her, he must marry her. Behind the other door is a ravenous tiger. As he chooses, he sees the princess point to the door on the right. He opens the door to find … well, there the story ends. To what fate did the princess send him?

Commentary

Physicists have been studying Newtonian mechanics ever since the great scientist published his book *Principia* in 1687. But some people still struggle with the book's content. In certain circumstances the laws of motion, as laid down by Newton, appear to contradict our everyday experience—and some people continue to favour their everyday experience over the findings of science. (This possibly explains why, even today, there are those who believe the Earth is flat: it *looks* flat so it must *be* flat.)

A famous example of a basic misunderstanding of Newton's laws of motion occurred in January 1920, when a *New York Times* editorial ridiculed the rocket pioneer Robert H. Goddard for daring to suggest a rocket might one day reach the Moon. The editorial stated:

> That Professor Goddard, with his 'chair' in Clark College and the countenancing of the Smithsonian Institution, does not know the relation of action to reaction, and of the need to have something better than a vacuum against which to react—to say that would be absurd. Of course he only seems to lack the knowledge ladled out daily in high schools.

The *New York Times* eventually published a good-natured, if belated, retraction. In July 1969, the day after the Apollo 11 rocket was on its way to the Moon, the editors wrote:

> Further investigation and experimentation have confirmed the findings of Isaac Newton in the 17th century and it is now definitely established that a rocket can function in a vacuum as well as in an atmosphere. The *Times* regrets the error.

In "A tale of negative gravity" Frank Stockton, who was writing in 1884, also gets Newtonian mechanics wrong—but he has more excuse than those nameless editors of the *New York Times*. It turns out negative gravity is a rather subtle concept.

*   *   *

When explaining how to overcome the force of gravity Stockton's inventor first presents some handwaving nonsense involving centrifugal force. Immediately afterwards, however, he defines what he means by negative gravity:

> Positive gravity attracts everything toward the centre of the earth. Negative gravity, therefore, would be that power which repels everything from the centre of the earth, just as the negative pole of a magnet repels the needle, while the positive pole attracts it.

This is the key notion: common sense suggests negative gravity would manifest itself as a repulsion, with two masses repelling each other just as two similar magnetic poles repel each other. If we could control the repulsion then we could float—as did the couple in Stockton's tale and the people illustrated in an 1878 *Punch* cartoon wearing antigravity underwear; see Fig. 9.1. (Perhaps Stockton got his idea for this story from reading *Punch*?) But it's not that simple.

Newton declared that any two masses experience a mutual gravitational force, the strength of which is directly proportional to the product of the masses and inversely proportional to the square of the distance between them. No variables other than mass and distance are involved. It's of no importance what the masses consist of, what colour they are, whether they are soft or hard—the gravitational force between two masses depends solely on the quantity of mass and the distance between the masses.

So imagine two masses, floating freely somewhere out in empty space. There will be an attractive gravitational force between them. If the distance between the masses *increases* by a factor of 2 then the force between them *decreases* by a factor of $2 \times 2 = 4$. If the distance between the masses *increases* by a factor of 3 then the force between them *decreases* by a factor of $3 \times 3 = 9$. And so on. Note, however, that the force is always one of *attraction*. The force depends upon the inverse *square* of the spatial separation between masses, so changing the distance between the masses changes the magnitude of the force but not its direction. (If the distance between the masses was made imaginary then the gravitational force between them would indeed change sign. But imaginary distance makes no physical sense. One can mutter the words, but the concept is meaningless.) So how can we change the direction of the force, and produce a repulsion rather than an attraction?

Well, a negative number multiplied by a positive number produces a negative number. So if one of the masses is negative and one is positive then the product of the masses will be negative. The force will be one of repulsion rather than attraction. And a repulsive force is what Stockton imagined when he described negative gravity.

Let's ignore, for the moment, the question of whether it's possible for a "negative mass" to exist and consider instead the question: would a negative mass behave in the way envisaged by Stockton in his story?

Suppose the inventor in Stockton's story was holding a mass of -1 kg. Then the *magnitude* of the gravitational force acting on the mass due to Earth's mass would be exactly the same as occurs on a mass of $+1$ kg; the *direction* of the force, however, would point away from the centre of the Earth. Repulsion rather than attraction. But how would that repulsion manifest itself? Well, Newton argued that a force acting on a mass causes the mass to accelerate. Earth's attractive gravitational force acting on a positive mass causes the mass

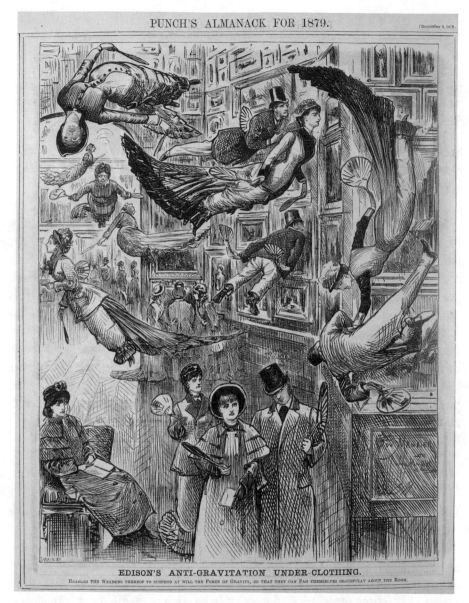

PUNCH'S ALMANACK FOR 1879. [December 9, 1878.

EDISON'S ANTI-GRAVITATION UNDER-CLOTHING.

ENABLES THE WEARERS THEREOF TO SUSPEND AT WILL THE FORCE OF GRAVITY, SO THAT THEY CAN FAN THEMSELVES GRACEFULLY ABOUT THE ROOM.

Fig. 9.1 This cartoon, one of several linked cartoons appearing in the 9 December 1878 issue of Punch, was drawn to illustrate an irreverent prediction of the inventions Edison might unveil in the forthcoming year (Credit: Public domain)

to accelerate towards the centre of the Earth. In other words, the mass falls. A *repulsive* force acting on a *negative* mass causes ... the mass to accelerate towards the centre of the Earth. In other words, the mass falls! The situation

Fig. 9.2   A sphere made of negative-mass material will repel a positive-mass spaceship, which will attract the negative-mass sphere ... which means the two objects accelerate away together (Credit: Author's own work)

Stockton describes involves two changes of sign: the direction of the gravitational force flips, but so does the response of the mass to an applied force. The effects cancel!

The effects of negative mass are far from intuitive.

Imagine you are the inventor in Stockton's story, and you have a kilogram of this negative mass material on the end of a string. You pull *down* on the string and feel a tug *upwards*, just as if you were holding a helium balloon. But if the string breaks the negative mass falls *down*, in exactly the same way as a positive mass falls. Or imagine a +1 kg mass and a −1 kg mass floating freely out in space. The negative mass experiences a negative gravitational force, but because of the way it responds to a force it accelerates towards the positive mass; the positive mass experiences a negative gravitational force and therefore accelerates *away* from the negative mass. Starting from rest the two masses self-accelerate along the line joining them, with the negative mass chasing the positive mass (see Fig. 9.2). This isn't what Stockton envisaged when he wrote about negative gravity, but it's what Newton's equations tell us will happen— if such a thing as negative mass exists.

* * *

Let's return to the question: is it possible to have a negative mass? Or is this two-word phrase as meaningless as "imaginary distance"?

Well, there's absolutely no evidence that any material in the universe possesses negative mass. Unlike the case of electric charge, which can be positive or negative, mass seems to come in only one variety: positive. That's why the forces of electromagnetism and gravity have such very different impacts on the universe. Electromagnetism is overwhelmingly more powerful than gravity, but when large amounts of matter are involved electric charges tend to cancel each other out. Gravity is a puny force but, because mass can pile upon mass without cancelling, the effects build up—with the result that, on the largest scales, gravity is the force that shapes the universe. If there were large amounts of negative matter in the universe then we'd soon know about it.

Are there any other hopes for antigravity? "What about antimatter?", I'm sure some people will say. "Doesn't antimatter possess negative mass?" Well, no, it doesn't. Compared to a normal particle, an antimatter particle does indeed have various charges reversed. The electron, for example, has electric charge −1 while its antimatter counterpart, the positron, has electric charge +1. However, although the electron and positron possess opposite electric charges, they possess the same mass: experiments show that they fall in Earth's gravity at the same rate. It's also worth mentioning that astronomers can point to evidence that matter and antimatter react in the same way to gravitational fields. SN1987A, a supernova in the Large Magellanic Cloud, created countless neutrinos and antineutrinos. These particles travelled 160,000 years to reach Earth, and the gravitational field of the Galaxy bent them from a straight-line path. This gravitationally-induced bending added five months to the travel time—and yet neutrinos and antineutrinos arrived at the same time. From this we can conclude that those cosmic neutrinos and antineutrinos "fell" in those gravitational wells at the same rate, to a precision of about one part in a million. Antimatter doesn't give us a route to antigravity.

What about relativity? We know Newton's theory of gravity has been superseded by Einstein's theory of general relativity. Does relativity permit antigravity?

In most cases Einstein's theory reduces to Newton's theory and so the Newtonian arguments against negative mass apply to relativity too. If it *were* possible to collect large amounts of negative mass then Einstein's theory opens up the possibility of exotic entities such as wormholes and warp drives. But the possibility seems remote.

The conclusion is almost inevitable: we are no more likely to achieve antigravity than we are to achieve faster then light travel. The universe just isn't set up in a way that would permit it.

* * *

The story in Chapter 10 examines a phenomenon that seems even more bizarre than antigravity—but that might, after a fashion, be possible.

Notes and Further Reading

a basic misunderstanding of Newton's laws of motion—The initial *New York Times* (1920) editorial was written in response to a popular article about rocketry that Goddard had published the previous year (Goddard 1919). He countered the editorial with an article published in *Scientific American* (Goddard 1921), but the damage had been done. Goddard retreated from the public eye. The *New York Times* (1969) retraction was beautifully dry, and didn't even mention the Apollo mission.

The effects of negative mass are far from intuitive—See, for example, Bondi (1957), Forward (1990), and Price (1993).

experiments show that they fall in Earth's gravity at the same rate—Note that the experiments necessary to test this statement are *incredibly* fiddly! At the time of writing, equipment is being installed at CERN that will permit physicists to measure directly the rate at which antiparticles fall under Earth's gravity (see ALPHA Collaboration and Charman (2013) for a description of the approach). The experiments need to be done without prejudice, of course, but no one expects the measurements will show any deviation between the rate at which matter and antimatter falls.

Bibliography

ALPHA Collaboration, Charman, A.E.: Description and first application of a new technique to measure the gravitational mass of antihydrogen. Nature Comm. **4**, 1785 (2013)

Bondi, H.: Negative mass in general relativity. Rev. Mod. Phys. **29**(3), 423–428 (1957)

Forward, R.L.: Negative matter propulsion. J. Prop. Power. **6**(1), 28–37 (1990)

Goddard, R.H.: A method of reaching extreme altitudes. Smithsonian Misc. Coll. **71**(2), 1–79 (1919)

Goddard, R.H.: That Moon rocket proposition: refutation of some popular fallacies. Sci. Am. 26 February (1921)

New York Times: Editorial. 13 January (1920)

New York Times: Editorial. 17 July (1969)

Price, R.H.: Negative mass can be positively amusing. Am. J. Phys. **63**, 216–217 (1993)

Stockton, F.R.: A tale of negative gravity. Century Mag. November issue (1884)

10

Matter Transmission

Prof. Vehr's Electrical Experiment (Robert Duncan Milne)

"The magic of the nineteenth century!" I exclaimed. "The term is so variously employed that it becomes necessary for me to know how you apply it before we can comprehend each other exactly."

"Well," responded Ashley, "I take it to mean the production of phenomena by natural means, which nevertheless seem supernatural, or beyond the present scope of applied science."

"As, for instance?" I inquired.

"Well, as, for instance, the faculty of intercommunication between persons separated by immense distances without the medium, say, of a tangible, physical telegraphic wire."

"But, in that case," I remarked, "and granting, for the sake of argument, that the intercommunication you describe might be carried on without the medium of a wire, how would you explain it?"

"By supposing," returned Ashley, "the existence of a real and actual medium whereby communications can be transmitted, though such medium is imperceptible to our ordinary senses, and cannot be weighed or measured by ordinary scientific instruments."

"But what reason have you," I objected, "for presuming the existence of any such medium at all?"

"The very best reason," he answered; "the actual experience of the past."

© Springer Nature Switzerland AG 2019
S. Webb, *New Light Through Old Windows: Exploring Contemporary Science Through 12 Classic Science Fiction Tales*, Science and Fiction,
https://doi.org/10.1007/978-3-030-03195-4_10

"You do not mean that you, personally, have communicated with—in short, transmitted messages to, and received messages from—distant persons without the use of ordinary telegraphic wires?" I asked.

"There is nothing so extraordinary in that. You must surely have witnessed the application of that law in the case of clairvoyants and trance mediums."

"Ah!" returned I. "But we were not talking of clairvoyants and trance mediums. Phenomena like these can be referred to a purely mental source. We were talking, I thought, of a real and actual medium, which could be proved to be such."

"Certainly; proved by results. Inferentially, that is to say, though the laws of its working still remain secret," replied my friend.

"I should like to witness such results myself," I said.

"You can do so by coming with me this evening," replied Ashley.

So it was agreed upon between the young physician, in whose room I then was, and myself, that we should meet again that evening at a certain spot, and afterward proceed to investigate the phenomena we had been talking about.

"And the fair Julia?" I remarked, inquiringly, changing the subject.

A shade passed over my friend's countenance as I made this remark. The lady I had referred to was his betrothed, and it required no great amount of observation on my part to perceive that in mentioning her name I had touched a tender cord, and so forebore to prosecute the subject, though entitled by intimacy with both to feel an interest in their mutual relations.

"It is this very matter which troubles me just now," replied my friend, uneasily. "Her letters have latterly been growing less frequent, and I fancy I can detect also a change of style. Her phrases seem less endearing than formerly. It would seem as though a shadow had sprung up between us. You know how I love her, and I am racked with apprehension when I think she is so far away, and exposed to I know not what estranging influences."

"Where are the family now?" I asked, as since the Radcliffes had left San Francisco, some six months before, they had, as I knew, been traveling in Europe, though I was unacquainted with their present location.

"That is what I do not know," replied Ashley, in tremulous tones. "They were in New York two weeks ago, and since then I have had no letter from Julia, though hitherto she has never missed a week without writing."

Julia Radcliffe, the affianced bride of Gerald Ashley, was a charming, sympathetic, and impressionable girl of nineteen summers, the only daughter of one of San Francisco's representative business men, who had been traveling with his family, and was now on his return home. Knowing the extent of my friend's affection for the lady, I felt sincere sympathy for him in his present condition.

Presently an idea struck me. Why should not my friend make use of the mode of communication he had just been explaining to me, and thus obtain the information he stood so sadly in need of? If it really possessed the virtue he expressed such confidence in, surely the present was the time to prove it, and I immediately made the suggestion.

"The very thing I had in my mind," he returned, in answer to my observation, "when I asked you to accompany me this evening. Professor Vehr is no ordinary scientist, I assure you. Some of the phenomena he produces are of the most extraordinary and startling character."

"Professor Vehr!" I exclaimed, in surprise. "Do you mean Professor Vehr of the Palace Hotel? I have already witnessed some of his remarkable experiments," recalling an episode in which a magic mirror had figured, some few months before. "I will willingly accompany you."

Two hours later found us in the professor's apartments in the Palace Hotel, where we were cordially welcomed.

"You have repeated your visit," he said, "with a view to further investigate the occult. Good. I shall be happy to oblige you—the more so as I have myself been pursuing the investigations, and have arrived at even more subtle elucidations of the energies conserved in the fluid we call electricity than those which you witnessed in the case of the mirror. There this phenomenon was confined to the reproduction of optical effects, with a certain reaction on the substantial forms of which the figures on the mirror were the *simulacra*. Now I am able to exert an actual physical control over the voices as well as the minds of distant persons themselves, so as, if necessary, to even transport them from place to place."

I was struck with the last observation of the professor, and with the resemblance of its claim to that of the adepts of the Oriental Theosophy, and so intimated.

"It is quite true," he said, "that this formulation of the energy I have just hinted at is, in effect, the same as that controlled by the Theosophists. It may be, also, that, in one sense, they have arrived at that more complete mastery over nature where the mere effort of mind and will is able to produce effects which I can only obtain in natural corollaries of laws which the world at large possesses. Still, even granting that such is the case, my mode of procedure for obtaining my results does not entail those penalties for their abuse to which explorers into the realms of the occult under purely physical conditions are exposed. There are, however, penalties equally terrible for a failure to observe the substantial conditions of scientific law—penalties which threaten absolute annihilation of individual identity, so far as the material person or *ego* can be annihilated"—and so saying, the professor turned toward the alcove which had been the scene of the experiment with the mirror.

It can readily be conceived that, while impressed with the deliberate enunciation and careful phraseology of the professor, I was somewhat at a loss to trace their application to a phenomenon which I had not yet witnessed, strongly tinctured as they were with that transcendental flavor which, while it whets the curiosity, tends rather to obscure than elucidate the subject on which it treats. I was not, therefore, sorry when I saw that our host was busying himself with the arrangement of some apparatus in the alcove, to which Ashley and myself now directed our attention, and waited.

The object which particularly arrested our attention was an immense glass bell, of something the shape and size of a diving-bell, which occupied the centre of the alcove. This bell rested, in an inverted position, upon a solid slab of plate-glass, touching its edges upon every side. It reminded one, shall I say, of the receiver of a gigantic airpump more than anything else. The capacity of this immense bell was, evidently, many hundred gallons, being, as nearly as I can judge, some five feet high by as many in diameter. Another peculiarity which I noted was that it was coated, to a height of about two feet from its base, with some shining opaque metallic substance; and as a metal rod descended from its apex about the same distance into its interior, while its upper end projected about a foot above the bell, I had no difficulty in connecting the apparatus before me with some branch of electricity, as the whole bore a marked resemblance to the known characteristics of a Leyden jar—that reservoir of static electricity whose scientific qualities are too well known to require further description. One other feature was deserving of notice—namely, that through one side of the bell projected the ends of what looked like ordinary telegraphic wires, terminating in those metal handles with which all who have experimented with the induction coil shock battery are familiar.

The professor walked leisurely round the bell, inspecting the metallic coating minutely, as likewise the wires which projected through the sides. Having apparently satisfied himself of the fitness of the apparatus before him, he turned to Ashley and said: "I received your letter regarding the lady, and shall be happy to assist you, so far as I can, in your search for information, and perhaps, in a still more hazardous experiment, if it becomes necessary, and if you have courage enough to undertake it. The first step, however, will be to ascertain the lady's present position and surroundings."

So saying the professor led the way to the glass bell, which he proceeded to raise from the floor, by means of a rope passing round a drum and through a pulley overhead. After raising it about four feet, he kept it in position by adjusting a ratchet on the drum, and, placing a chair upon the glass slab, requested Ashley to seat himself thereon. Ashley did so, and the professor then put into each of his hands one of the handles terminating the wires which ran through the side of the bell. These wires I now saw ran to the outer wall of the apartment, where they disappeared.

"They are," explained the professor, "merely private connections with our ordinary public telegraph wires. They are, however, under certain conditions, rendered peculiarly sensitive to currents which would be powerless to affect ordinary telegraphic instruments in any manner whatever." Thus saying he loosened the ratchet on the drum, and proceeded to let the glass bell descend till it completely covered Ashley, enclosing him as if in a vase, its edges resting on the glass slab on which his chair was set.

"There is, as you see," explained the professor, in answer, as it were, to an unspoken idea, "no danger of asphyxiation, as the holes through which the side wires and the top rod pass, are by no means tight, this condition not being essential to the success of the experiment."

He then proceeded to take cautiously from a glass jar at one side of the alcove, the end of a piece of thick rubber tubing, a wire at the end of which he attached to a hook on the metallic coating of a bell; and then, from a second glass jar in another corner, a similar piece of tubing, the end of which, by standing on a chair, he connected with the end of the rod which projected upward from the apex of the bell. I noticed that these latter india-rubber-coated wires ran out into the street like the simple telegraph wires first mentioned. I noticed, too, that they looked peculiarly similar to the rubber-coated wires used for conveying the fluid which feeds the electric lights in streets and public buildings.

"It is now," said the professor, descending from his chair, after attaching the second wire to the rod at the top of the bell, "that the nicety of our experiment comes in. I have had these insulated wires, from the works of the Electric Lighting Company, brought into my apartment in order to save time and trouble in charging my bell, an operation tedious to accomplish by generating electricity by a plate apparatus in the ordinary manner. Still, the great speed and force with which the fluid is generated and transmitted through these wires necessitates extreme care in manipulation. A degree too much in tension might be productive of the most serious consequences to anyone confined inside. Still, there is no fear in the first stage of the experiment, and none, if strict care is taken, in the second." So saying, the professor approached the bell with an electrometer, and presently, after disconnecting the rubber-coated wires therefrom, returning them carefully to their respective insulators.

It was now apparent that while we were conversing Ashley had closed his eyes, and was now reclining back in his chair, seemingly in deep sleep, his hands still clasping the handles of the telegraph-wires that ran into the interior of the bell.

"That is the first effect of a moderately strong charge of static electricity in the human frame," explained the professor. "It induces a highly wrought condition of the nerves, which in their turn act upon the ganglion of the brain; that, in its turn, reacting again, through the duplex series of nerves, upon the wire held in the left hand, which brings the holder into communication with whatever object enthrals his attention at the time of trance. The experiment is, in effect, clairvoyance reduced to an art, the mesmeric trance accomplished by scientific means and conditioned by the recognized and accepted laws of electrical science. Your friend is now, as I verily believe, in direct spiritual communication with her who is dearest to his heart—the last object that held possession of his soul before the mesmeric-electric trance overtook him. I will put myself in communication with him and ascertain."

The professor then walked around to the side of the bell where the wires entered the glass, and applied the forefinger of either hand to the orifices. As he did so, Ashley gave a visible start. His eyes, however, still remained closed. "Have you seen her?" asked the professor, in deliberate tones, bending his spectacled eyes upon Ashley. "Where is she? What do you see?"

"I see a vast building—a collection of vast buildings. I see throngs of people, gayly dressed, walking in and out of them, and parading the beautiful grounds which surround them. It reminds me of the Centennial Exposition of '76."

"He is evidently in New Orleans," observed the professor to me. "I have no question that Miss Radcliffe is there. Do you see the lady?" he continued, addressing my friend.

"Ha! There!" responded Ashley, with animation. "There is Julia with her father and mother, and stay!—there is another there—a tall, handsome man, who has just approached the party. He takes off his hat and bows. He steps to Julia's side. She shrinks a little as he approaches. Mr. and Mrs. Radcliffe smile as he proffers his arm. They walk on. He is bending over Julia and conversing in low, passionate tones. He is telling her that he loves her. She listens to him as in a dream. He presses his suit more vehemently. They have left or lost her father and mother in the crowd. They are now alone in a little Moorish kiosk. He bends down and kisses her, and though she shudders she does not move away. Merciful heavens!" and with a start Gerald Ashley awoke.

The professor eyed him calmly. "Are you satisfied?" he said.

Great beads of perspiration were standing on Ashley's brow. He looked the picture of agony and apprehension.

"You have been there in spirit," said the professor. "You know and appreciate the position your love is in. Would you regain her? Would you save her to yourself? Do you dare to appear beside her in bodily form, and bring her back with you here, wresting her from the new lover who has gained dominion over her, and whom, if you hesitate now, you will be in a very short time powerless to rival? Do you trust to my art? Believe me, I sympathize with you, and will help you if I can," and the professor looked calmly and steadily into the eyes of the man inside the bell.

"I am willing to encounter any risk," responded Ashley, "to regain the love of my betrothed. What must I do?"

"Simply retain your clasp upon the handles," said the professor, calmly. "Keep your attention fixed, as heretofore, and you will presently be in the kiosk of the New Orleans Exposition, not in spirit as you were a minute ago, but in actual, physical body. When you find yourself there, I leave it to yourself to regain the affections of your betrothed. Leave it to me to bring you both here. Be careful, though, not to loosen your clasp on her hand when the wire sounds the call to return. Go, and fortune attend you."

Ashley again clasped the handles of the wires with a set determination in his face which betokened that he realized and appreciated every detail of the professor's advice. The latter walked with somewhat quicker step than was his wont to the insulators where the wires of the Electric Lighting Company lay, and proceeded to adjust them to their respective places on the bell as before. Presently Ashley's head fell back upon the chair as before, and his eyes closed. The professor took out his watch, looked at it somewhat nervously, approached the bell at intervals with his electrometer, and paced the floor.

"My instrument is based," he explained to me, "on a centigrade scale of my own. It will take some time for a bell of the size you see before you to become fully charged with the fluid, even with my wires, which I have had conveyed almost straight from headquarters; and until the bell is fully charged the experiment cannot be consummated."

We waited patiently for some minutes more, Ashley murmuring meanwhile: "New Orleans; kiosk; I can see her; I can see him, but I can not approach her; her father and mother are searching anxiously for her through other portions of the grounds."

Presently my friend became silent. The professor again, for the fifth or sixth time, approached the bell and applied his instrument.

"Hush!" said he. "The tension in the bell is now augmenting rapidly. A short time more and we shall witness the successful accomplishment of our experiment."

As he spoke, I noticed a change taking place in Ashley. Seated there, as he was, his head leaning on the back of his chair, his hands grasping the electrodes, I distinctly saw his form become thin, filmy, and transparent. Moment by moment more thin, filmy, and transparent it grew, till the attenuation was such that even the outline was scarcely visible. The hands alone, of all portions of the body, retained something of their pristine substantiality.

"He has gone," whispered the professor, with subdued excitement. "We must now gauge the time till we can reasonably presume that he has accomplished his purpose, regaining his betrothed, and take measures for their return—the return of both, for if he be without her little would be gained. She would again, doubtless, as soon as her lover has left her, become amenable to him who gained a dominion over her in the absence of her original lover, and whose supremacy would be restored."

"But how—how—" I stammered "how is this to be accounted for? What has become of my friend who was seated there but a moment ago?"

"The simplest thing in the world, my dear sir," replied the professor, earnestly. "You desire to know why your friend's body has disappeared. Would you ask the same question had that body been exposed to intense heat? You answer, no; because you say that it is a recognized law of chemistry that the most refractory substances—rock or metal—are first melted and then volatilized by heat. Heat, in effect, expands and disintegrates the mass, and decollocates the atoms of every known substance. The static electricity, with which that glass bell is charged, is, in one sense, a correlative of heat, and, in another, a correlative of the physical energy which we call spirit. Is it wonderful, therefore, that the intense force with which that bell is laden should suffice to produce the effects which you now witness? There is nothing wonderful, my dear sir, in any chemical process, no matter how apparently incomprehensible and inexplicable it may be. You have seen a solid block of ice made instantaneously in and projected from a red-hot crucible. Is what is now being done any more incomprehensible than that?

"Say that the human body you lately saw before you has been volatilized, resolved into its primary elements, and then, through the agency of the psychic power resident within it, has been transmitted to the spot which that psychic power had most in view, along the ordinary telegraph wires which connect us with that spot. Sound waves can be transmitted in this manner by the telephone; light waves are amenable to the same law—why, then, should not this law apply to matter which has been etherealized to the same degree as light and sound? It is but carrying the subject to its legitimate conclusion. But stay! Hush! Here they come!"

As the professor spoke, I became aware of a dim and shadowy form shaping itself within the bell. As it gained shape I discerned, with what feelings of awe can be imagined, that there was not *one* form but *two*. Slowly but gradually the dim form assumed a more bodily and distinct aspect I came in a very few seconds to realize that the forms which stood before me in the bell were those of Gerald Ashley and his affianced bride, Julia Radcliffe. There could be no doubt about the matter. There they stood in distinct bodily form, but still with a bewildered, dreamy look upon them, as though scarcely awakened from sleep.

"We have triumphed!" said the professor, in low tones, as he mounted on a chair to disconnect the insulated feeding wire from the rod at the top of the bell, at the same time motioning me to do the same for the other wire. I proceeded to do so as fast as I could, when suddenly a blinding flash passed before my eyes, and a report sounded in my ears like that of a cannon. The last thing I remember seeing was the great glass bell with Gerald Ashley and Julia Radcliffe inside, and Professor Vehr standing on a chair beside it, the whole picture illumined by the brilliant light of an electric flash, and the *facade* of Market Street distinctly visible through the windows of the apartment.

* * *

A week afterward, when I recovered sensibility, I ascertained several things. First, that Professor Vehr had left the city, the lessee of the Palace Hotel vowing that he would never receive another scientist, or permit another private wire from the Electric Lighting Company to enter the building. Secondly, that Doctor Gerald Ashley had not been seen since the eventful night, the strange experiences of which I have just recorded. Thirdly, that on that very day Miss Julia Radcliffe had disappeared from New Orleans, the last time that she had been seen being in a certain kiosk in the Exposition grounds with a Mr. Arthur Livingstone, though that gentleman avers that he saw her approached by some stranger, who claimed an acquaintanceship with her, and with whom she disappeared in the crowd a short time afterward. When Mr. Livingstone was pressed for a description of the gentleman in question, Mr. and Mrs. Radcliffe were at once convinced that it could be no one else but Doctor Gerald Ashley; and while sadly recriminating him for such a hasty and unadvised step as eloping with their daughter, have no doubt that the pair will turn up when it becomes expedient to do so.

As for me, I have taken pains to verify the date of the disappearance of Miss Julia Radcliffe at New Orleans, and find it to be the same day and hour as Dr. Gerald Ashley's and my own visit to Professor Vehr's apartments, on the night of the grand explosion at the Palace Hotel. I have arrived at a certain conclusion on the facts, and consider that the most judicious as well as the most sympathetic course I can pursue, under the circumstances, is to place the foregoing facts before the public just as they are, thereby, perhaps, affording that poor solace to the afflicted which consists in the removal of an ill-grounded hope of ever seeing their lost and loved ones again.

Robert Duncan Milne (1844–1899) was a Scottish-born journalist and author who moved to California sometime in the 1860s. Beyond that, there isn't a huge amount of biographical detail about him. Although Milne was well known at the time, and although one could make a claim for him being the first full-time science fiction writer in America (he had five dozen or so SF stories published in the San Francisco literary journal *The Argonaut*) he became largely forgotten after his unfortunate death (Milne was struck by one of San Francisco's cable cars).

Commentary

Most stories in this book require modern readers to suspend their knowledge of science and technology. This is particularly the case with "Prof. Vehr's electrical experiment". The author, Robert Duncan Milne, has a character suggest there's nothing extraordinary about the transmission of information in the absence of wires to carry the information: after all, "clairvoyants and trance mediums" seem to do this all the time. (He was not alone in his blithe acceptance of clairvoyance. Arthur Conan Doyle may have been a more famous writer than Milne but he was also more gullible: Doyle attended séances and made countless public speeches to promote his belief in ghosts, mediums, and the spirit world.) For modern readers, of course, the wireless transmission of information is entirely unremarkable—our daily lives depend upon it—but the notion that clairvoyance is a routine phenomenon, merely a form of mental WiFi, strikes us as foolishness. So how could Milne get it so wrong?

Milne was writing in 1885. For his readers, the transmission of information over wire was well established. The first electrical telegraph systems had been developed in the 1830s, and by the 1870s many parts of the world had been linked by cable. Science fiction writers are in the business of extrapolation, so it was natural to want to describe a technology enabling long-distance communication *without* wires. The easiest way to frame such a technology was by invoking clairvoyance—Milne's readers would certainly have encountered tales of "clairvoyants and trance mediums" and, even if they dismissed such ideas as nonsense, they would have understood what Milne was getting at. Had Milne sought to frame wireless communication using the physics available to him then his readers would undoubtedly have been confused.

The notion that electromagnetic waves can propagate freely through space—and thus permit wireless transmission of information—was developed mathematically by James Clerk Maxwell in 1864. Few if any of Milne's readers would have been able to appreciate the implications of Maxwell's work, however, and the existence of radio waves was demonstrated experimentally only after the publication of Milne's story. It wasn't until 1886 that Heinrich Hertz began a series of experiments that proved the reality of electromagnetic waves; and radio waves would not form the basis of a commercial wireless telegraphy system until the turn of the century, when Marconi began investigating the technology. The first practical *television* systems, which could transmit live moving images, were not developed until the 1920s—four decades or so after Milne's story. So we can reasonably cut Milne some slack in this regard: what we take for granted is a relatively modern advance.

In his story, however, Milne goes far beyond what our radio and television technology permits us to do. Clairvoyancy is supposed to permit not only the wireless transmission of information but also the wireless transmission of *matter*. And matter transmission, or teleportation, is impossible—isn't it?

*   *   *

Teleportation is a science fiction staple. The most famous example of a matter transmitter is surely *Star Trek*'s Transporter—a wonderful plot device that meant Captain Kirk had only to utter the phrase "Beam me up, Scotty" before being whisked to safety—but matter transmission formed the basis of numerous SF stories. Although writers in the Golden Age of science fiction might have doubted the possibility of a practical teleportation system, they were in a better position than Milne to provide a rationale for the phenomenon. Radio and TV sets relied upon the wireless transmission of *information*, so SF readers could accept the following hand-waving explanation of matter transmission. A matter transmitter would "scan" an object (to obtain full information about the material of which it was made), transmit that information to a receiving station, and then use the information to reconstitute the object. Simple, no?

No one suggested it would be easy to make a matter transmitter but the recipe outlined above—"scan" an object at point A; transmit full information about the object to point B; use that information to reconstitute the object—sounds as if it should work in principle. Put this way, teleportation sounds like an extremely advanced engineering problem. But then physicists took a look, and pointed out that quantum mechanics spoils this wonderful science fictional dream of matter transmission.

The difficulty is the Uncertainty Principle. This fundamental tenet of quantum mechanics says you can't simultaneously measure the position and momentum of a particle. But that means you can't get full information about an object when you "scan" it. This lack of complete information ensures the object recreated at point B can't be a completely faithful copy of the original. In the worst case, the recreated object might end up being undifferentiated mush—a prospect that worried *Star Trek*'s Dr McCoy whenever the medic was forced to use the Transporter.

So physicists had shown how a fundamental principle of quantum mechanics ruled out a working teleportation device. "Shame", I hear SF readers say. But then, in 1993, Charles Bennett and colleagues showed how a different fundamental principle of quantum mechanics does indeed permit a form of teleportation! So—what's going on?

* * *

The key to quantum teleportation is the mysterious phenomenon of quantum entanglement.

Entanglement occurs when a pair of particles interact in such a way that the quantum state of one particle can't be described independently of the other—even if the particles are widely separated. (Figure 10.1 is an artist's attempt to represent a concept that is perhaps incapable of being represented.) Entangled particles possess properties—spin, polarization, and so on—that are correlated. For example, suppose we create a pair of entangled particles with a total spin of zero. Then, if we measure the spin of one particle to be "up" then the spin of the other particle will be measured to be "down"—no matter how widely separated the two particles happen to be. The mysterious aspect of this phenomenon arises because, until a measurement is made, each of the quantum particles will be in a superposition of spin-up and spin-down states. To put it crudely, the particle hasn't decided whether it's spinning up or down until we observe it. Making a measurement *here* causes the state of the particle here to be spin-up (or spin-down, depending on what we observe); but the

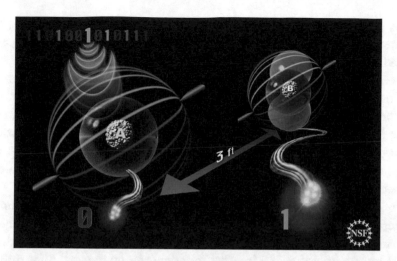

Fig. 10.1 An artist's attempt to represent a phenomenon which, although it has been demonstrated in the laboratory, is almost impossible to visualise. The goal of quantum teleportation is to transfer the quantum state of particle A onto a distant particle B—which is possible if the particles are entangled. When the goal is achieved it's information, not matter, that's transmitted. In everyday language, however, the net effect is the same as transmitting a quantum particle from A to B. An artist created this image to illustrate an experiment, carried out in 2009, that teleported an atomic state over a distance of one metre, or about 3 feet: the state of ion A on the left got scrambled but was reconstructed at ion B on the right (Credit: Nicolle Rager Fuller, NSF)

measurement *instantaneously* causes the state of the other particle over *there* to be spin-down (or spin-up, depending on what we observe here). And that instantaneous effect happens regardless of the size of the separation. The particles could be on different continents or on different planets—entanglement still holds.

Entanglement seems paradoxical but experiments demonstrate its validity. Entanglement may be a mystery but we may as well learn to live with it. How, though, does entanglement get round the Uncertainty Principle and help with teleportation?

Let's keep things simple, and agree to teleport a single particle. Alice is going to send the particle and Bob is going to receive it.

Alice starts by entangling two particles—let's call them A and B—and she keeps particle A while sending particle B to Bob. This sets up the transmission channel: if Alice performs some operation on particle A then *instantaneously* this affects the state of particle B. Now, suppose Alice has a particle C whose quantum state she wants to teleport to Bob. To do this she makes a particular type of quantum measurement on particles A and C. The act of measuring erases the state from C: the particle C no longer exists in the original form. However, the entanglement between A and B means B is in a state that can be turned into whatever state C had—so long as Bob makes the correct operation on B. In order for Bob to know what operation to apply, Alice has to send him the outcome of her measurement—and she has to do this by a classical channel (she could phone him, send him a postcard, email him, or whatever). Once Bob has this information he can turn B into a state that's identical to C. And because the state of a quantum particle tells you all there is to know about the particle—all that it's *possible* to know about the particle—Alice has succeeded in teleporting particle C to Bob.

It's important to understand that Alice can't use this recipe to transport matter or energy; what gets transferred is a quantum state. Furthermore, Alice can't use this recipe to transmit information faster than light: entanglement might be instantaneous, but in order for teleportation to work Alice must also transmit classical information, which can only happen at light speed or slower. Nevertheless, Alice *can* use the recipe to teleport a quantum state. The recipe might seem complicated, but physicists have demonstrated it successfully in the laboratory.

In 1998, just five years after publication of the initial protocol outlining how quantum teleportation might work, physicists succeeded in teleporting a single photon over a small distance. Six years after that, they succeeded in teleporting a photon over a distance of 600 m. Then they learned how to teleport electrons, atoms, and superconducting circuits. At the time of writing, the record distance for teleporting a quantum system is 1400 km. So quantum teleportation is now well established. Quantum teleportation couldn't explain the events in Milne's story, nor can it be used to construct a *Star Trek* Transporter. But it *will* form the basis of usable technologies—just not the technologies SF writers once dreamed of.

*   *   *

Clairvoyance also plays a role in the next story. When it came to séance-based mumbo-jumbo Fitz-James O'Brien, the next story's author, was as credulous as Milne and Doyle. It's a shame these authors were unaware of quantum theory—as we'll see in Chapter 11, this theory gives us our best understanding of the world on the smallest scales.

Notes and Further Reading

Doyle attended séances—Doyle has been the subject of numerous biographies, some of which focus on his life beyond the creation of Sherlock Holmes and Professor Challenger. See, for example, Stashower (1999); Polidoro (2001); Miller (2008); Fox (2018).

wireless transmission of information—Lewis (1991) is a good account of the development of radio, with a particular emphasis on American pioneers; Burns (1998) gives the early history of television.

Star Trek's *Transporter*—For more details of the Transporter, along with photon torpedoes, visors, and more, see Siegel (2017).

Charles Bennett and colleagues—The foundational paper is Bennett et al. (1993).

Alice starts by entangling two particles—Entanglement will provide the basis of several new technologies, including quantum computing. Aaronson (2013) discusses the deep principles behind these emerging technologies. Aaronson's book is funny, too.

Bibliography

Aaronson, S.: Quantum Computing Since Democritus. Cambridge University Press, Cambridge (2013)

Bennett, C.H., Brassard, G., Crépeau, C., Jozsa, R., Peres, A., Wootters, W.K.: Teleporting an unknown quantum state via dual classical and Einstein–Podolsky–Rosen channels. Phys. Rev. Lett. **70**(13), 1895–1899 (1993)

Burns, R.W.: Television: An International History of the Formative Years. Institution of Engineering and Technology, Stevenage (1998)

Fox, M.: Conan Doyle for the Defense: The True Story of a Sensational British Murder, a Quest for Justice, and the World's Most Famous Detective Writer. Random House, New York (2018)

Lewis, T.: Empire of the Air: The Men Who Made Radio. HarperCollins, New York (1991)

Miller, R.: The Adventures of Arthur Conan Doyle: A Biography. Thomas Dunne, New York (2008)

Milne, R.D.: Prof. Vehr's electrical experiment. Argonaut. January issue (1885)

Polidoro, M.: Final Séance: The Strange Friendship Between Houdini and Conan Doyle. Prometheus, New York (2001)

Siegel, E.: Treknology: The Science of Star Trek from Tricorders to Warp Drive. Voyageur, Minneapolis, MN (2017)

Stashower, D.: Teller of Tales: The Life of Arthur Conan Doyle. Henry Holt, New York (1999)

11

The Sub-Microscopic World

The Diamond Lens (Fitz-James O'Brien)

I

FROM a very early period of my life the entire bent of my inclinations had been toward microscopic investigations. When I was not more than ten years old, a distant relative of our family, hoping to astonish my inexperience, constructed a simple microscope for me by drilling in a disk of copper a small hole in which a drop of pure water was sustained by capillary attraction. This very primitive apparatus, magnifying some fifty diameters, presented, it is true, only indistinct and imperfect forms, but still sufficiently wonderful to work up my imagination to a preternatural state of excitement.

Seeing me so interested in this rude instrument, my cousin explained to me all that he knew about the principles of the microscope, related to me a few of the wonders which had been accomplished through its agency, and ended by promising to send me one regularly constructed, immediately on his return to the city. I counted the days, the hours, the minutes that intervened between that promise and his departure.

Meantime, I was not idle. Every transparent substance that bore the remotest resemblance to a lens I eagerly seized upon, and employed in vain attempts to realize that instrument the theory of whose construction I as yet only vaguely comprehended. All panes of glass containing those oblate spheroidal

© Springer Nature Switzerland AG 2019
S. Webb, *New Light Through Old Windows: Exploring Contemporary Science Through 12 Classic Science Fiction Tales*, Science and Fiction,
https://doi.org/10.1007/978-3-030-03195-4_11

knots familiarly known as "bull's-eyes" were ruthlessly destroyed in the hope of obtaining lenses of marvelous power. I even went so far as to extract the crystalline humor from the eyes of fishes and animals, and endeavored to press it into the microscopic service. I plead guilty to having stolen the glasses from my Aunt Agatha's spectacles, with a dim idea of grinding them into lenses of wondrous magnifying properties—in which attempt it is scarcely necessary to say that I totally failed.

At last the promised instrument came. It was of that order known as Field's simple microscope, and had cost perhaps about fifteen dollars. As far as educational purposes went, a better apparatus could not have been selected. Accompanying it was a small treatise on the microscope—its history, uses, and discoveries. I comprehended then for the first time the "Arabian Nights' Entertainments." The dull veil of ordinary existence that hung across the world seemed suddenly to roll away, and to lay bare a land of enchantments. I felt toward my companions as the seer might feel toward the ordinary masses of men. I held conversations with nature in a tongue which they could not understand. I was in daily communication with living wonders such as they never imagined in their wildest visions. I penetrated beyond the external portal of things, and roamed through the sanctuaries. Where they beheld only a drop of rain slowly rolling down the window-glass, I saw a universe of beings animated with all the passions common to physical life, and convulsing their minute sphere with struggles as fierce and protracted as those of men. In the common spots of mould, which my mother, good housekeeper that she was, fiercely scooped away from her jam-pots, there abode for me, under the name of mildew, enchanted gardens, filled with dells and avenues of the densest foliage and most astonishing verdure, while from the fantastic boughs of these microscopic forests hung strange fruits glittering with green and silver and gold.

It was no scientific thirst that at this time filled my mind. It was the pure enjoyment of a poet to whom a world of wonders has been disclosed. I talked of my solitary pleasures to none. Alone with my microscope, I dimmed my sight, day after day and night after night, poring over the marvels which it unfolded to me. I was like one who, having discovered the ancient Eden still existing in all its primitive glory, should resolve to enjoy it in solitude, and never betray to mortals the secret of its locality. The rod of my life was bent at this moment. I destined myself to be a microscopist.

Of course, like every novice, I fancied myself a discoverer. I was ignorant at the time of the thousands of acute intellects engaged in the same pursuit as myself, and with the advantage of instruments a thousand times more powerful than mine. The names of Leeuwenhoek, Williamson, Spencer, Ehrenberg,

Schultz, Dujardin, Schact, and Schleiden were then entirely unknown to me, or, if known, I was ignorant of their patient and wonderful researches. In every fresh specimen of cryptogamia which I placed beneath my instrument I believed that I discovered wonders of which the world was as yet ignorant. I remember well the thrill of delight and admiration that shot through me the first time that I discovered the common wheel animalcule (*Rotifera vulgaris*) expanding and contracting its flexible spokes and seemingly rotating through the water. Alas! as I grew older, and obtained some works treating of my favorite study, I found that I was only on the threshold of a science to the investigation of which some of the greatest men of the age were devoting their lives and intellects.

As I grew up, my parents, who saw but little likelihood of anything practical resulting from the examination of bits of moss and drops of water through a brass tube and a piece of glass, were anxious that I should choose a profession.

It was their desire that I should enter the counting-house of my uncle, Ethan Blake, a prosperous merchant, who carried on business in New York. This suggestion I decisively combated. I had no taste for trade; I should only make a failure; in short, I refused to become a merchant.

But it was necessary for me to select some pursuit. My parents were staid New England people, who insisted on the necessity of labor, and therefore, although, thanks to the bequest of my poor Aunt Agatha, I should, on coming of age, inherit a small fortune sufficient to place me above want, it was decided that, instead of waiting for this, I should act the nobler part, and employ the intervening years in rendering myself independent.

After much cogitation, I complied with the wishes of my family, and selected a profession. I determined to study medicine at the New York Academy. This disposition of my future suited me. A removal from my relatives would enable me to dispose of my time as I pleased without fear of detection. As long as I paid my Academy fees, I might shirk attending the lectures if I chose; and, as I never had the remotest intention of standing an examination, there was no danger of my being "plucked." Besides, a metropolis was the place for me. There I could obtain excellent instruments, the newest publications, intimacy with men of pursuits kindred with my own—in short, all things necessary to ensure a profitable devotion of my life to my beloved science. I had an abundance of money, few desires that were not bounded by my illuminating mirror on one side and my object-glass on the other; what, therefore, was to prevent my becoming an illustrious investigator of the veiled worlds? It was with the most buoyant hope that I left my New England home and established myself in New York.

II

My first step, of course, was to find suitable apartments. These I obtained, after a couple of days' search, in Fourth Avenue; a very pretty second floor, unfurnished, containing sitting-room, bedroom, and a smaller apartment which I intended to fit up as a laboratory. I furnished my lodgings simply, but rather elegantly, and then devoted all my energies to the adornment of the temple of my worship. I visited Pike, the celebrated optician, and passed in review his splendid collection of microscopes—Field's Compound, Hingham's, Spencer's, Nachet's Binocular (that founded on the principles of the stereo-scope), and at length fixed upon that form known as Spencer's Trunnion Microscope, as combining the greatest number of improvements with an almost perfect freedom from tremor. Along with this I purchased every pos-sible accessory—draw-tubes, micrometers, a *camera lucida*, lever-stage, achro-matic condensers, white cloud illuminators, prisms, parabolic condensers, polarizing apparatus, forceps, aquatic boxes, fishing-tubes, with a host of other articles, all of which would have been useful in the hands of an experi-enced microscopist, but, as I afterward discovered, were not of the slightest present value to me. It takes years of practice to know how to use a compli-cated microscope. The optician looked suspiciously at me as I made these valuable purchases. He evidently was uncertain whether to set me down as some scientific celebrity or a madman. I think he was inclined to the latter belief. I suppose I was mad. Every great genius is mad upon the subject in which he is greatest. The unsuccessful madman is disgraced and called a lunatic.

Mad or not, I set myself to work with a zeal which few scientific students have ever equaled. I had everything to learn relative to the delicate study upon which I had embarked—a study involving the most earnest patience, the most rigid analytic powers, the steadiest hand, the most untiring eye, the most refined and subtle manipulation.

For a long time half my apparatus lay inactively on the shelves of my labo-ratory, which was now most amply furnished with every possible contrivance for facilitating my investigations. The fact was that I did not know how to use some of my scientific implements—never having been taught microscopies—and those whose use I understood theoretically were of little avail until by practice I could attain the necessary delicacy of handling. Still, such was the fury of my ambition, such the untiring perseverance of my experiments, that, difficult of credit as it may be, in the course of one year I became theoretically and practically an accomplished microscopist.

During this period of my labors, in which I submitted specimens of every substance that came under my observation to the action of my lenses, I became a discoverer—in a small way, it is true, for I was very young, but still a discoverer. It was I who destroyed Ehrenberg's theory that the *Volvox globator* was an animal, and proved that his "monads" with stomachs and eyes were merely phases of the formation of a vegetable cell, and were, when they reached their mature state, incapable of the act of conjugation, or any true generative act, without which no organism rising to any stage of life higher than vegetable can be said to be complete. It was I who resolved the singular problem of rotation in the cells and hairs of plants into ciliary attraction, in spite of the assertions of Wenham and others that my explanation was the result of an optical illusion.

But notwithstanding these discoveries, laboriously and painfully made as they were, I felt horribly dissatisfied. At every step I found myself stopped by the imperfections of my instruments. Like all active microscopists, I gave my imagination full play. Indeed, it is a common complaint against many such that they supply the defects of their instruments with the creations of their brains. I imagined depths beyond depths in nature which the limited power of my lenses prohibited me from exploring. I lay awake at night constructing imaginary microscopes of immeasurable power, with which I seemed to pierce through all the envelopes of matter down to its original atom. How I cursed those imperfect mediums which necessity through ignorance compelled me to use! How I longed to discover the secret of some perfect lens, whose magnifying power should be limited only by the resolvability of the object, and which at the same time should be free from spherical and chromatic aberrations—in short, from all the obstacles over which the poor microscopist finds himself continually stumbling! I felt convinced that the simple microscope, composed of a single lens of such vast yet perfect power, was possible of construction. To attempt to bring the compound microscope up to such a pitch would have been commencing at the wrong end; this latter being simply a partially successful endeavor to remedy those very defects of the simplest instrument which, if conquered, would leave nothing to be desired.

It was in this mood of mind that I became a constructive microscopist. After another year passed in this new pursuit, experimenting on every imaginable substance—glass, gems, flints, crystals, artificial crystals formed of the alloy of various vitreous materials—in short, having constructed as many varieties of lenses as Argus had eyes—I found myself precisely where I started, with nothing gained save an extensive knowledge of glass-making. I was almost dead with despair. My parents were surprised at my apparent want of progress in my medical studies (I had not attended one lecture since my arrival in the city), and the expenses of my mad pursuit had been so great as to embarrass me very seriously.

I was in this frame of mind one day, experimenting in my laboratory on a small diamond—that stone, from its great refracting power, having always occupied my attention more than any other—when a young Frenchman who lived on the floor above me, and who was in the habit of occasionally visiting me, entered the room.

I think that Jules Simon was a Jew. He had many traits of the Hebrew character: a love of jewelry, of dress, and of good living. There was something mysterious about him. He always had something to sell, and yet went into excellent society. When I say sell, I should perhaps have said peddle; for his operations were generally confined to the disposal of single articles—a picture, for instance, or a rare carving in ivory, or a pair of duelling-pistols, or the dress of a Mexican *caballero*. When I was first furnishing my rooms, he paid me a visit, which ended in my purchasing an antique silver lamp, which he assured me was a Cellini—it was handsome enough even for that—and some other knick-knacks for my sitting-room. Why Simon should pursue this petty trade I never could imagine. He apparently had plenty of money, and had the *entrée* of the best houses in the city—taking care, however, I suppose, to drive no bargains within the enchanted circle of the Upper Ten. I came at length to the conclusion that this peddling was but a mask to cover some greater object, and even went so far as to believe my young acquaintance to be implicated in the slave-trade. That, however, was none of my affair.

On the present occasion, Simon entered my room in a state of considerable excitement.

"*Ah! mon ami!*" he cried, before I could even offer him the ordinary salutation, "it has occurred to me to be the witness of the most astonishing things in the world. I promenade myself to the house of Madame ————. How does the little animal—*le renard*—name himself in the Latin?"

"Vulpes," I answered.

"Ah! yes—Vulpes. I promenade myself to the house of Madame Vulpes."

"The spirit medium?"

"Yes, the great medium. Great heavens! what a woman! I write on a slip of paper many of questions concerning affairs of the most secret—affairs that conceal themselves in the abysses of my heart the most profound; and behold, by example, what occurs? This devil of a woman makes me replies the most truthful to all of them. She talks to me of things that I do not love to talk of to myself. What am I to think? I am fixed to the earth!"

"Am I to understand you, M. Simon, that this Mrs. Vulpes replied to questions secretly written by you, which questions related to events known only to yourself?"

"Ah! more than that, more than that," he answered, with an air of some alarm. "She related to me things—But," he added after a pause, and suddenly changing his manner, "why occupy ourselves with these follies? It was all the biology, without doubt. It goes without saying that it has not my credence. But why are we here, *mon ami*? It has occurred to me to discover the most beautiful thing as you can imagine—a vase with green lizards on it, composed by the great Bernard Palissy. It is in my apartment; let us mount. I go to show it to you."

I followed Simon mechanically; but my thoughts were far from Palissy and his enameled ware, although I, like him, was seeking in the dark a great discovery. This casual mention of the spiritualist, Madame Vulpes, set me on a new track. What if, through communication with more subtle organisms than my own, I could reach at a single bound the goal which perhaps a life, of agonizing mental toil would never enable me to attain?

While purchasing the Palissy vase from my friend Simon, I was mentally arranging a visit to Madame Vulpes.

III

Two evenings after this, thanks to an arrangement by letter and the promise of an ample fee, I found Madame Vulpes awaiting me at her residence alone. She was a coarse-featured woman, with keen and rather cruel dark eyes, and an exceedingly sensual expression about her mouth and under jaw. She received me in perfect silence, in an apartment on the ground floor, very sparsely furnished. In the centre of the room, close to where Mrs. Vulpes sat, there was a common round mahogany table. If I had come for the purpose of sweeping her chimney, the woman could not have looked more indifferent to my appearance. There was no attempt to inspire the visitor with awe. Everything bore a simple and practical aspect. This intercourse with the spiritual world was evidently as familiar an occupation with Mrs. Vulpes as eating her dinner or riding in an omnibus.

"You come for a communication, Mr. Linley?" said the medium, in a dry, businesslike tone of voice.

"By appointment—yes."

"What sort of communication do you want—a written one?"

"Yes, I wish for a written one."

"From any particular spirit?"

"Yes."

"Have you ever known this spirit on this earth?"

"Never. He died long before I was born. I wish merely to obtain from him some information which he ought to be able to give better than any other."

"Will you seat yourself at the table, Mr. Linley," said the medium, "and place your hands upon it?"

I obeyed, Mrs. Vulpes being seated opposite to me, with her hands also on the table. We remained thus for about a minute and a half, when a violent succession of raps came on the table, on the back of my chair, on the floor immediately under my feet, and even on the window-panes. Mrs. Vulpes smiled composedly.

"They are very strong to-night," she remarked. "You are fortunate." She then continued, "Will the spirits communicate with this gentleman?"

Vigorous affirmative.

"Will the particular spirit he desires to speak with communicate?" A very confused rapping followed this question.

"I know what they mean," said Mrs. Vulpes, addressing herself to me; "they wish you to write down the name of the particular spirit that you desire to converse with. Is that so?" she added, speaking to her invisible guests.

That it was so was evident from the numerous affirmatory responses. While this was going on, I tore a slip from my pocket-book and scribbled a name under the table.

"Will this spirit communicate in writing with this gentleman?" asked the medium once more.

After a moment's pause, her hand seemed to be seized with a violent tremor, shaking so forcibly that the table vibrated. She said that a spirit had seized her hand and would write. I handed her some sheets of paper that were on the table and a pencil. The latter she held loosely in her hand, which presently began to move over the paper with a singular and seemingly involuntary motion. After a few moments had elapsed, she handed me the paper, on which I found written, in a large, uncultivated hand, the words, "He is not here, but has been sent for." A pause of a minute or so ensued, during which Mrs. Vulpes remained perfectly silent, but the raps continued at regular intervals. When the short period I mention had elapsed, the hand of the medium was again seized with its convulsive tremor, and she wrote, under this strange influence, a few words on the paper, which she handed to me. They were as follows:

"I am here. Question me."

"*Leeuwenhoek.*"

I was astounded. The name was identical with that I had written beneath the table, and carefully kept concealed. Neither was it at all probable that an uncultivated woman like Mrs. Vulpes should know even the name of the great father of microscopies. It may have been biology; but this theory was soon doomed to be destroyed. I wrote on my slip—still concealing it from Mrs. Vulpes—a series of questions which, to avoid tediousness, I shall place with the responses, in the order in which they occurred:

I.—Can the microscope be brought to perfection?

Spirit—Yes.

I.—Am I destined to accomplish this great task?

Spirit.—You are.

I.—I wish to know how to proceed to attain this end. For the love which you bear to science, help me!

Spirit—A diamond of one hundred and forty carats, submitted to electro-magnetic currents for a long period, will experience a rearrangement of its atoms *inter se* and from that stone you will form the universal lens.

I.—Will great discoveries result from the use of such a lens?

Spirit—So great that all that has gone before is as nothing.

I.—But the refractive power of the diamond is so immense that the image will be formed within the lens. How is that difficulty to be surmounted?

Spirit—Pierce the lens through its axis, and the difficulty is obviated. The image will be formed in the pierced space, which will itself serve as a tube to look through. Now I am called. Good-night.

I can not at all describe the effect that these extraordinary communications had upon me. I felt completely bewildered. No biological theory could account for the *discovery* of the lens. The medium might, by means of biological *rapport* with my mind, have gone so far as to read my questions and reply to them coherently. But biology could not enable her to discover that magnetic currents would so alter the crystals of the diamond as to remedy its previous defects and admit of its being polished into a perfect lens. Some such theory may have passed through my head, it is true; but if so, I had forgotten it. In my excited condition of mind there was no course left but to become a convert, and it was in a state of the most painful nervous exaltation that I left the medium's house that evening. She accompanied me to the door, hoping that I was satisfied. The raps followed us as we went through the hall, sounding on the balusters, the flooring, and even the lintels of the door. I hastily expressed my satisfaction, and escaped hurriedly into the cool night air. I walked home with but one thought possessing me—how to obtain a diamond of the immense size required. My entire means multiplied a hundred times over would have been inadequate to its purchase. Besides, such stones are rare, and become historical. I could find such only in the regalia of Eastern or European monarchs.

IV

There was a light in Simon's room as I entered my house. A vague impulse urged me to visit him. As I opened the door of his sitting-room unannounced, he was bending, with his back toward me, over a Carcel lamp, apparently engaged in minutely examining some object which he held in his hands. As I entered, he started suddenly, thrust his hand into his breast pocket, and turned to me with a face crimson with confusion.

"What!" I cried, "poring over the miniature of some fair lady? Well, don't blush so much; I won't ask to see it."

Simon laughed awkwardly enough, but made none of the negative protestations usual on such occasions. He asked me to take a seat.

"Simon," said I, "I have just come from Madame Vulpes."

This time Simon turned as white as a sheet, and seemed stupefied, as if a sudden electric shock had smitten him. He babbled some incoherent words, and went hastily to a small closet where he usually kept his liquors. Although astonished at his emotion, I was too preoccupied with my own idea to pay much attention to anything else.

"You say truly when you call Madame Vulpes a devil of a woman," I continued. "Simon, she told me wonderful things to-night, or rather was the means of telling me wonderful things. Ah! if I could only get a diamond that weighed one hundred and forty carats!"

Scarcely had the sigh with which I uttered this desire died upon my lips when Simon, with the aspect of a wild beast, glared at me savagely, and, rushing to the mantelpiece, where some foreign weapons hung on the wall, caught up a Malay creese, and brandished it furiously before him.

"No!" he cried in French, into which he always broke when excited. "No! you shall not have it! You are perfidious! You have consulted with that demon, and desire my treasure! But I will die first! Me, I am brave! You can not make me fear!"

All this, uttered in a loud voice, trembling with excitement, astounded me. I saw at a glance that I had accidentally trodden upon the edges of Simon's secret, whatever it was. It was necessary to reassure him.

"My dear Simon," I said, "I am entirely at a loss to know what you mean. I went to Madame Vulpes to consult with her on a scientific problem, to the solution of which I discovered that a diamond of the size I just mentioned was necessary. You were never alluded to during the evening, nor, so far as I was concerned, even thought of. What can be the meaning of this outburst? If you happen to have a set of valuable diamonds in your possession, you need fear nothing from me. The diamond which I require you could not possess; or, if you did possess it, you would not be living here."

Something in my tone must have completely reassured him, for his expression immediately changed to a sort of constrained merriment, combined however, with a certain suspicious attention to my movements. He laughed, and said that I must bear with him; that he was at certain moments subject to a species of vertigo, which betrayed itself in incoherent speeches, and that the attacks passed off as rapidly as they came.

He put his weapon aside while making this explanation, and endeavored, with some success, to assume a more cheerful air.

All this did not impose on me in the least. I was too much accustomed to analytical labors to be baffled by so flimsy a veil. I determined to probe the mystery to the bottom.

"Simon," I said gayly, "let us forget all this over a bottle of Burgundy. I have a case of Lausseure's *Clos Vougeot* downstairs, fragrant with the odors and ruddy with the sunlight of the Côte d'Or. Let us have up a couple of bottles. What say you?"

"With all my heart," answered Simon smilingly.

I produced the wine and we seated ourselves to drink. It was of a famous vintage, that of 1848, a year when war and wine throve together, and its pure but powerful juice seemed to impart renewed vitality to the system. By the time we had half finished the second bottle, Simon's head, which I knew was a weak one, had begun to yield, while I remained calm as ever, only that every draught seemed to send a flush of vigor through my limbs. Simon's utterance became more and more indistinct. He took to singing French *chansons* of a not very moral tendency. I rose suddenly from the table just at the conclusion of one of those incoherent verses, and, fixing my eyes on him with a quiet smile, said, "Simon, I have deceived you. I learned your secret this evening. You may as well be frank with me. Mrs. Vulpes—or rather, one of her spirits—told me all."

He started with horror. His intoxication seemed for the moment to fade away, and he made a movement toward the weapon that he had a short time before laid down, I stopped him with my hand.

"Monster!" he cried passionately, "I am ruined! What shall I do? You shall never have it! I swear by my mother!"

"I don't want it," I said; "rest secure, but be frank with me. Tell me all about it."

The drunkenness began to return. He protested with maudlin earnestness that I was entirely mistaken—that I was intoxicated; then asked me to swear eternal secrecy, and promised to disclose the mystery to me. I pledged myself, of course, to all. With an uneasy look in his eyes, and hands unsteady with drink and nervousness, he drew a small case from his breast and opened it.

Heavens! How the mild lamplight was shivered into a thousand prismatic arrows as it fell upon a vast rose-diamond that glittered in the case! I was no judge of diamonds, but I saw at a glance that this was a gem of rare size and purity. I looked at Simon with wonder and—must I confess it?—with envy. How could he have obtained this treasure? In reply to my questions, I could just gather from his drunken statements (of which, I fancy, half the incoherence was affected) that he had been superintending a gang of slaves engaged in diamond-washing in Brazil; that he had seen one of them secrete a diamond, but, instead of informing his employers, had quietly watched the negro until he saw him bury his treasure; that he had dug it up and fled with it, but that as yet he was afraid to attempt to dispose of it publicly—so valuable a gem being almost certain to attract too much attention to its owner's antecedents—and he had not been able to discover any of those obscure channels by which such matters are conveyed away safely. He added that, in accordance with oriental practice, he had named his diamond with the fanciful title of "The Eye of Morning."

While Simon was relating this to me, I regarded the great diamond attentively. Never had I beheld anything so beautiful. All the glories of light ever imagined or described seemed to pulsate in its crystalline chambers. Its weight, as I learned from Simon, was exactly one hundred and forty carats. Here was an amazing coincidence. The hand of destiny seemed in it. On the very evening when the spirit of Leeuwenhoek communicates to me the great secret of the microscope, the priceless means which he directs me to employ start up within my easy reach! I determined, with the most perfect deliberation, to possess myself of Simon's diamond.

I sat opposite to him while he nodded over his glass, and calmly revolved the whole affair. I did not for an instant contemplate so foolish an act as a common theft, which would of course be discovered, or at least necessitate flight and concealment, all of which must interfere with my scientific plans. There was but one step to be taken—to kill Simon. After all, what was the life of a little peddling Jew in comparison with the interests of science? Human beings are taken every day from the condemned prisons to be experimented on by surgeons. This man, Simon, was by his own confession a criminal, a robber, and I believed on my soul a murderer. He deserved death quite as much as any felon condemned by the laws: why should I not, like government, contrive that his punishment should contribute to the progress of human knowledge?

The means for accomplishing everything I desired lay within my reach. There stood upon the mantelpiece a bottle half full of French laudanum. Simon was so occupied with his diamond, which I had just restored to him, that it was an affair of no difficulty to drug his glass. In a quarter of an hour he was in a profound sleep.

I now opened his waistcoat, took the diamond from the inner pocket in which he had placed it, and removed him to the bed, on which I laid him so that his feet hung down over the edge. I had possessed myself of the Malay creese, which I held in my right hand, while with the other I discovered as accurately as I could by pulsation the exact locality of the heart. It was essential that all the aspects of his death should lead to the surmise of self-murder. I calculated the exact angle at which it was probable that the weapon, if leveled by Simon's own hand, would enter his breast; then with one powerful blow I thrust it up to the hilt in the very spot which I desired to penetrate. A convulsive thrill ran through Simon's limbs. I heard a smothered sound issue from his throat, precisely like the bursting of a large air-bubble sent up by a diver when it reaches the surface of the water; he turned half round on his side, and, as if to assist my plans more effectually, his right hand, moved by some mere spasmodic impulse, clasped the handle of the creese, which it remained holding with extraordinary muscular tenacity. Beyond this there was no apparent struggle. The laudanum, I presume, paralyzed the usual nervous action. He must have died instantly.

There was yet something to be done. To make it certain that all suspicion of the act should be diverted from any inhabitant of the house to Simon himself, it was necessary that the door should be found in the morning *locked on the in-side*. How to do this, and afterward escape myself? Not by the window; that was a physical impossibility. Besides, I was determined that the windows also should be found bolted. The solution was simple enough. I descended softly to my own room for a peculiar instrument which I had used for holding small slippery substances, such as minute spheres of glass, etc. This instrument was nothing more than a long, slender hand-vise, with a very powerful grip and a considerable leverage, which last was accidentally owing to the shape of the handle. Nothing was simpler than, when the key was in the lock, to seize the end of its stem in this vise, through the keyhole, from the outside, and so lock the door. Previously, however, to doing this, I burned a number of papers on Simon's hearth. Suicides almost always burn papers before they destroy themselves. I also emptied some more laudanum into Simon's glass—having first removed from it all traces of wine—cleaned the other wine-glass, and brought the bottles away with me. If traces of two persons drinking had been found in the room, the question naturally would have arisen, Who was the second? Besides, the wine-bottles might have been identified as belonging to me. The laudanum I poured out to account for its presence in his stomach, in case of a *post-mortem* examination. The theory naturally would be that he first intended to poison himself, but, after swallowing a little of the drug, was either disgusted with its taste, or changed his mind from other motives, and chose the dagger. These arrangements made, I walked out, leaving the gas burning, locked the door with my vise, and went to bed.

Simon's death was not discovered until nearly three in the afternoon. The servant, astonished at seeing the gas burning—the light streaming on the dark landing from under the door—peeped through the keyhole and saw Simon on the bed.

She gave the alarm. The door was burst open, and the neighborhood was in a fever of excitement.

Every one in the house was arrested, myself included. There was an inquest; but no clew to his death beyond that of suicide could be obtained. Curiously enough, he had made several speeches to his friends the preceding week that seemed to point to self-destruction. One gentleman swore that Simon had said in his presence that "he was tired of life." His landlord affirmed that Simon, when paying him his last month's rent, remarked that "he should not pay him rent much longer." All the other evidence corresponded—the door locked inside, the position of the corpse, the burned papers. As I anticipated, no one knew of the possession of the diamond by Simon, so that no motive was suggested for his murder. The jury, after a prolonged examination, brought in the usual verdict, and the neighborhood once more settled down to its accustomed quiet.

V

The three months succeeding Simon's catastrophe I devoted night and day to my diamond lens. I had constructed a vast galvanic battery, composed of nearly two thousand pairs of plates: a higher power I dared not use, lest the diamond should be calcined. By means of this enormous engine I was enabled to send a powerful current of electricity continually through my great diamond, which it seemed to me gained in lustre every day. At the expiration of a month I commenced the grinding and polishing of the lens, a work of intense toil and exquisite delicacy. The great density of the stone, and the care required to be taken with the curvatures of the surfaces of the lens, rendered the labor the severest and most harassing that I had yet undergone. At last the eventful moment came; the lens was completed. I stood trembling on the threshold of new worlds. I had the realization of Alexander's famous wish before me. The lens lay on the table, ready to be placed upon its platform. My hand fairly shook as I enveloped a drop of water with a thin coating of oil of turpentine, preparatory to its examination, a process necessary in order to prevent the rapid evaporation of the water. I now placed the drop on a thin slip of glass under the lens, and throwing upon it, by the combined aid of a prism and a mirror, a powerful stream of light, I approached my eye to the minute hole drilled through the axis of the lens. For an instant I saw nothing save what seemed to be an illuminated chaos, a vast, luminous abyss. A pure white light, cloudless and serene, and seemingly limitless as space itself, was my first impression. Gently, and with the greatest care, I depressed the lens a few hairbreadths. The wondrous illumination still continued, but as the lens approached the object a scene of indescribable beauty was unfolded to my view.

I seemed to gaze upon a vast space, the limits of which extended far beyond my vision. An atmosphere of magical luminousness permeated the entire field of view. I was amazed to see no trace of animalculous life. Not a living thing, apparently, inhabited that dazzling expanse. I comprehended instantly that, by the wondrous power of my lens, I had penetrated beyond the grosser particles of aqueous matter, beyond the realms of infusoria and protozoa, down to the original gaseous globule, into whose luminous interior I was gazing as into an almost boundless dome filled with a supernatural radiance. It was, however, no brilliant void into which I looked. On every side I beheld beautiful inorganic forms, of unknown texture, and colored with the most enchanting hues. These forms presented the appearance of what might be called, for want of a more specific definition, foliated clouds of the highest rarity—that is, they undulated and broke into vegetable formations, and were tinged with splendors compared with which the gilding of our autumn woodlands is as dross compared with gold. Far away into the illimitable distance stretched long avenues of these gaseous forests, dimly transparent, and painted with prismatic hues of unimaginable brilliancy. The pendent branches waved along the fluid glades until every vista seemed to break through half-lucent ranks of many-colored drooping silken pennons. What seemed to be either fruits or flowers, pied with a thousand hues, lustrous and ever-varying, bubbled from the crowns of this fairy foliage. No hills, no lakes, no rivers, no forms animate or inanimate, were to be seen, save those vast auroral copses that floated serenely in the luminous stillness, with leaves and fruits and flowers gleaming with unknown fires, unrealizable by mere imagination.

How strange, I thought, that this sphere should be thus condemned to solitude! I had hoped, at least, to discover some new form of animal life, perhaps of a lower class than any with which we are at present acquainted, but still some living organism. I found my newly discovered world, if I may so speak, a beautiful chromatic desert.

While I was speculating on the singular arrangements of the internal economy of Nature, with which she so frequently splinters into atoms our most compact theories, I thought I beheld a form moving slowly through the glades of one of the prismatic forests. I looked more attentively, and found that I was not mistaken. Words can not depict the anxiety with which I awaited the nearer approach of this mysterious object. Was it merely some inanimate substance, held in suspense in the attenuated atmosphere of the globule, or was it an animal endowed with vitality and motion? It approached, flitting behind the gauzy, colored veils of cloud-foliage, for seconds dimly revealed, then vanishing. At last the violet pennons that trailed nearest to me vibrated; they were gently pushed aside, and the form floated out into the broad light. It was a female human shape. When I say human, I mean it possessed the outlines of humanity; but there the analogy ends. Its adorable beauty lifted it illimitable heights beyond the loveliest daughter of Adam.

I can not, I dare not, attempt to inventory the charms of this divine revelation of perfect beauty. Those eyes of mystic violet, dewy and serene, evade my words. Her long, lustrous hair following her glorious head in a golden wake, like the track sown in heaven by a falling star, seems to quench my most burning phrases with its splendors. If all the bees of Hybla nestled upon my lips, they would still sing but hoarsely the wondrous harmonies of outline that inclosed her form.

She swept out from between the rainbow-curtains of the cloud-trees into the broad sea of light that lay beyond. Her motions were those of some graceful naiad, cleaving, by a mere effort of her will, the clear, unruffled waters that fill the chambers of the sea. She floated forth with the serene grace of a frail bubble ascending through the still atmosphere of a June day. The perfect roundness of her limbs formed suave and enchanting curves. It was like listening to the most spiritual symphony of Beethoven the divine, to watch the harmonious flow of lines. This, indeed, was a pleasure cheaply purchased at any price. What cared I if I had waded to the portal of this wonder through another's blood. I would have given my own to enjoy one such moment of intoxication and delight.

Breathless with gazing on this lovely wonder, and forgetful for an instant of everything save her presence, I withdrew my eye from the microscope eagerly. Alas! as my gaze fell on the thin slide that lay beneath my instrument, the bright light from mirror and from prism sparkled on a colorless drop of water! There, in that tiny bead of dew, this beautiful being was forever imprisoned. The planet Neptune was not more distant from me than she. I hastened once more to apply my eye to the microscope.

Animula (let me now call her by that dear name which I subsequently bestowed on her) had changed her position. She had again approached the wondrous forest, and was gazing earnestly upward. Presently one of the trees—as I must call them—unfolded a long ciliary process, with which it seized one of the gleaming fruits that glittered on its summit, and, sweeping slowly down, held it within reach of Animula. The sylph took it in her delicate hand and began to eat. My attention was so entirely absorbed by her that I could not apply myself to the task of determining whether this singular plant was or was not instinct with volition.

I watched her, as she made her repast, with the most profound attention. The suppleness of her motions sent a thrill of delight through my frame; my heart beat madly as she turned her beautiful eyes in the direction of the spot in which I stood. What would I not have given to have had the power to pre-cipitate myself into that luminous ocean and float with her through those grooves of purple and gold! While I was thus breathlessly following her every movement, she suddenly started, seemed to listen for a moment, and then cleaving the brilliant ether in which she was floating, like a flash of light, pierced through the opaline forest and disappeared.

Instantly a series of the most singular sensations attacked me. It seemed as if I had suddenly gone blind. The luminous sphere was still before me, but my daylight had vanished. What caused this sudden disappearance? Had she a lover or a husband? Yes, that was the solution! Some signal from a happy fellow-being had vibrated through the avenues of the forest, and she had obeyed the summons.

The agony of my sensations, as I arrived at this conclusion, startled me. I tried to reject the conviction that my reason forced upon me. I battled against the fatal conclusion—but in vain. It was so. I had no escape from it. I loved an animalcule.

It is true that, thanks to the marvelous power of my microscope, she appeared of human proportions. Instead of presenting the revolting aspect of the coarser creatures, that live and struggle and die, in the more easily resolv-able portions of the water-drop, she was fair and delicate and of surpassing beauty. But of what account was all that? Every time that my eye was with-drawn from the instrument it fell on a miserable drop of water, within which, I must be content to know, dwelt all that could make my life lovely.

Could she but see me once! Could I for one moment pierce the mystical walls that so inexorably rose to separate us, and whisper all that filled my soul, I might consent to be satisfied for the rest of my life with the knowledge of her remote sympathy.

It would be something to have established even the faintest personal link to bind us together—to know that at times, when roaming through these enchanted glades, she might think of the wonderful stranger who had broken the monotony of her life with his presence and left a gentle memory in her heart!

But it could not be. No invention of which human intellect was capable could break down the barriers that nature had erected. I might feast my soul upon her wondrous beauty, yet she must always remain ignorant of the adoring eyes that day and night gazed upon her, and, even when closed, beheld her in dreams. With a bitter cry of anguish I fled from the room, and flinging myself on my bed, sobbed myself to sleep like a child.

VI

I arose the next morning almost at daybreak, and rushed to my microscope, I trembled as I sought the luminous world in miniature that contained my all. Animula was there. I had left the gas-lamp, surrounded by its moderators, burning when I went to bed the night before. I found the sylph bathing, as it were, with an expression of pleasure animating her features, in the brilliant light which surrounded her. She tossed her lustrous golden hair over her shoulders with innocent coquetry. She lay at full length in the transparent medium, in which she supported herself with ease, and gamboled with the enchanting grace that the nymph Salmacis might have exhibited when she sought to conquer the modest Hermaphroditus. I tried an experiment to satisfy myself if her powers of reflection were developed. I lessened the lamplight considerably. By the dim light that remained, I could see an expression of pain flit across her face. She looked upward suddenly, and her brows contracted. I flooded the stage of the microscope again with a full stream of light, and her whole expression changed. She sprang forward like some substance deprived of all weight. Her eyes sparkled and her lips moved. Ah! if science had only the means of conducting and reduplicating sounds, as it does rays of light, what carols of happiness would then have entranced my ears! what jubilant hymns to Adonais would have thrilled the illumined air!

I now comprehended how it was that the Count de Cabalis peopled his mystic world with sylphs—beautiful beings whose breath of life was lambent fire, and who sported forever in regions of purest ether and purest light. The Rosicrucian had anticipated the wonder that I had practically realized.

How long this worship of my strange divinity went on thus I scarcely know. I lost all note of time. All day from early dawn, and far into the night, I was to be found peering through that wonderful lens. I saw no one, went nowhere, and scarce allowed myself sufficient time for my meals. My whole life was absorbed in contemplation as rapt as that of any of the Romish saints. Every hour that I gazed upon the divine form strengthened my passion—a passion that was always overshadowed by the maddening conviction that, although I could gaze on her at will, she never, never could behold me!

At length I grew so pale and emaciated, from want of rest and continual brooding over my insane love and its cruel conditions, that I determined to make some effort to wean myself from it. "Come," I said, "this is at best but a fantasy. Your imagination has bestowed on Animula charms which in reality she does not possess. Seclusion from female society has produced this morbid condition of mind. Compare her with the beautiful women of your own world, and this false enchantment will vanish."

I looked over the newspapers by chance. There I beheld the advertisement of a celebrated *danseuse* who appeared nightly at Niblo's. The Signorina Caradolce had the reputation of being the most beautiful as well as the most graceful woman in the world. I instantly dressed and went to the theatre.

The curtain drew up. The usual semicircle of fairies in white muslin were standing on the right toe around the enameled flower-bank of green canvas, on which the belated prince was sleeping. Suddenly a flute is heard. The fairies start. The trees open, the fairies all stand on the left toe, and the queen enters. It was the Signorina. She bounded forward amid thunders of applause, and, lighting on one foot, remained poised in the air. Heavens! was this the great enchantress that had drawn monarchs at her chariot-wheels? Those heavy, muscular limbs, those thick ankles, those cavernous eyes, that stereotyped smile, those crudely painted cheeks! Where were the vermeil blooms, the liquid, expressive eyes, the harmonious limbs of Animula?

The Signorina danced. What gross, discordant movements! The play of her limbs was all false and artificial. Her bounds were painful athletic efforts; her poses were angular and distressed the eye. I could bear it no longer; with an exclamation of disgust that drew every eye upon me, I rose from my seat in the very middle of the Signorina's *pas-de-fascination* and abruptly quitted the house.

I hastened home to feast my eyes once more on the lovely form of my sylph. I felt that henceforth to combat this passion would be impossible. I applied my eyes to the lens. Animula was there—but what could have happened? Some terrible change seemed to have taken place during my absence. Some secret grief seemed to cloud the lovely features of her I gazed upon. Her face had grown thin and haggard; her limbs trailed heavily; the wondrous lustre of her golden hair had faded. She was ill—ill, and I could not assist her! I believe at that moment I would have forfeited all claims to my human birthright if I could only have been dwarfed to the size of an animalcule, and permitted to console her from whom fate had forever divided me.

I racked my brain for the solution of this mystery. What was it that afflicted the sylph? She seemed to suffer intense pain. Her features contracted, and she even writhed, as if with some internal agony. The wondrous forests appeared also to have lost half their beauty. Their hues were dim and in some places faded away altogether. I watched Animula for hours with a breaking heart, and she seemed absolutely to wither away under my very eye. Suddenly I remembered that I had not looked at the water-drop for several days. In fact, I hated to see it; for it reminded me of the natural barrier between Animula and myself. I hurriedly looked down on the stage of the microscope. The slide was still there—but, great heavens, the water drop had vanished! The awful truth burst upon me; it had evaporated, until it had become so minute as to be invisible to the naked eye; I had been gazing on its last atom, the one that contained Animula—and she was dying!

I rushed again to the front of the lens and looked through. Alas! the last agony had seized her. The rainbow-hued forests had all melted away, and Animula lay struggling feebly in what seemed to be a spot of dim light. Ah! the sight was horrible: the limbs once so round and lovely shriveling up into nothings; the eyes—those eyes that shone like heaven—being quenched into black dust; the lustrous golden hair now lank and discolored. The last throe came. I beheld that final struggle of the blackening form—and I fainted.

When I awoke out of a trance of many hours, I found myself lying amid the wreck of my instrument, myself as shattered in mind and body as it. I crawled feebly to my bed, from which I did not rise for many months.

They say now that I am mad; but they are mistaken. I am poor, for I have neither the heart nor the will to work; all my money is spent, and I live on charity. Young men's associations that love a joke invite me to lecture on optics before them, for which they pay me, and laugh at me while I lecture. "Linley, the mad microscopist," is the name I go by. I suppose that I talk incoherently while I lecture. Who could talk sense when his brain is haunted by such ghastly memories, while ever and anon among the shapes of death I behold the radiant form of my lost Animula!

Fitz-James O'Brien (1826–1862) was born as Michael O'Brien in Cork, Ireland, but in around 1852 changed his name to Fitz-James and emigrated to America. He wrote a number of poems and stories, including an early science fiction tale about invisibility, and was an established author by the time the American Civil War began. O'Brien enlisted into the New York National Guard, and was wounded in battle—leading to an infection that killed him.

Commentary

Fitz-James O'Brien, writing in 1858, was clearly as familiar as Robert Milne (see previous chapter) with the activities of spirit mediums. However, if we ignore the paranormal aspects of this story (as well as the offensive nature of the narrator) it is clear O'Brien had a passing acquaintance with the science—or at least the microscopy—of his day. With the obvious exception of the diamond lens itself, the instruments he describes were all working microscopes and the names he drops were all noted microscopists.

For example, Antonie Philips van Leeuwenhoek, whose voice plays a role in the story, is widely considered to be the first microscopist. He used a primitive sort of device—essentially just a small glass ball set in a metal frame—to discover "animalcules" (what we now call microorganisms). Christian Gottfried Ehrenberg, another noted microscopist mentioned in the story, spent three decades of his scientific career examining water, soil, and dust under magnification and he described countless new species. Félix Dujardin was an equally accomplished microscopist, as was Matthias Jakob Schleiden, who studied the structure of plants and was the co-founder of one of the most important principles of biology—namely, that organisms are composed of fundamental units called cells. And Charles Achilles Spencer was America's first microscope maker to meet with success—his "Trunnion Microscope", which O'Brien's narrator purchases, was classed as one of the finest optical instruments of the mid-19th century.

Those early microscopes, although relatively simple, were advanced enough to provide sufficient magnification for the science of microbiology to develop. It turns out you don't need a huge amount of magnification to observe various types of specimen. At a magnification of 40×, for example, you can see the larger types of biological cell. At a magnification of 100× you can make out bacteria as tiny dots. Increase the magnification to 400× and bacterial shapes become clearly visible while individual chromosomes can be seen within cells. So—given the state of microscopy in the mid-1850s, was O'Brien justified in imagining the miniature forest world inhabited by Animula? I'm not convinced he was.

*   *   *

Even Victorian instruments were sufficiently sophisticated to cast doubt on the notion that the microscopic world is like the everyday world, only smaller. To be sure, in the early days of microscopy Ehrenberg had argued that animalcules were "complete organisms". In other words, there was an early view that all animals—from microscopic organisms through to elephants—possess complete organ systems: muscular system, circulatory system, lymphatic system, and so on. But as early as 1835, microscopic studies carried by Dujardin had disproved this idea. So, by the time O'Brien was writing, the world inhabited by Animula was more fantasy than science fiction.

Can we be entirely sure, however, that going to higher magnifications won't uncover a hidden world of tiny people inhabiting tiny forests? Well, yes we can.

Consider how a miniature human such as Animula might function. Better still, think about one of those many other miniaturized humans that appear in science fiction from time to time: Scott Carey, the protagonist in Richard Matheson's *The Shrinking Man*, provides perhaps the most famous example. Towards the end of Matheson's novel, Carey concludes that he won't simply disappear as he continues to shrink. He thinks: "If nature existed on endless levels, so also might intelligence." Carey's belief was the general philosophy behind all these stories, from O'Brien's "The Diamond Lens" through to Matheson's *The Shrinking Man* almost a century later. But how could an intelligent being such as Animula or the ever-dwindling Scott Carey possibly exist? The human brain contains about 100 billion neurons. If the miniature person must maintain the mass of all those neurons—not to mention all the other systems required by the human form—then he or she is going to have the density of a white dwarf star! This is simply not viable. As the person gets smaller, the only way to preserve the density that's typical of biological organisms is to get rid of mass. But if Animula or Carey had 100 neurons, say, rather than 100 billion neurons … well, they'd be less like a human and more like a rotifer (one of the tiny animals seen by microscopists such as van Leeuwenhoek, Ehrenberg, and Dujardin; see Fig. 11.1). The venerable SF trope of humans in microscopic form has never made any sense.

The possibility of advanced life forms existing at the microscopic level is ruled out by observation as well as theory.

Nowadays, microbiologists have access to optical microscopes capable of magnifying up to about 1500×. (Beyond this magnification the object being viewed starts to appear fuzzy; the wavelength of light puts a limit on the clarity of images.) This is sufficient magnification to make out bacteria and the various constituents of a biological cell. And it's clear that the microbial world does not contain creatures of the complexity of Animula. Unicellular organisms, yes; complex animals, no.

What about the world at even shorter distances? Might intricate structures exist at those length scales? Well, modern techniques permit microscopists to

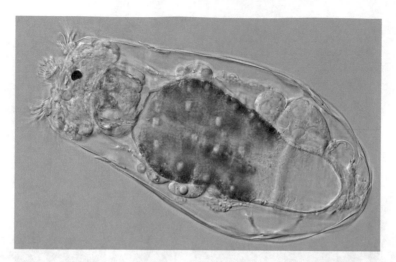

Fig. 11.1   A rotifer of the genus Northolca. Most rotifers are between 0.1–0.5 mm in size, and so they are among the smallest animals; you need a microscope to study them. Many of the first rotifers to be described were identified by van Leeuwenhoek at the start of the 18th century. But van Leeuwenhoek was working when the techniques of optical microscopy were in their infancy; the beautiful image here, which clearly shows various body parts inside a shell-like protective outer covering, could only have been taken using modern techniques. Clearly, in an animal as small as this, there simply isn't enough room to contain a nervous system that would support the behaviour exhibited by Animula in "The Diamond Lens" (Credit: Wiedehopf20)

apply magnifications greatly beyond the limit of traditional optical microscopes and, again, observations rule out Animula.

In 1931, the German physicist Ernst Ruska developed the first transmission electron microscope (TEM). The TEM is similar in function to an optical microscope but uses electrons rather than light; where an optical microscope bends light rays with a glass lens, a TEM bends electron beams with an electromagnet. A modern TEM allows scientists to produce extremely high-resolution images at extremely high magnifications—thousands of times greater than optical telescopes permit. And four years after the invention of the TEM, Ruska's colleague Max Knoll developed the scanning electron microscope (SEM). A SEM doesn't have the same extremely high resolution of a TEM, but it has numerous other advantages: those spectacular photographs we see of the surface of microscopic creatures are usually SEM images (see, for example, Fig. 11.2). More recently microscopists have developed other techniques in addition to TEM and SEM. None of these approaches or techniques, when applied to biological systems, have found anything other than cells and the usual structures that are to be found in cells. The world of Animula doesn't exist.

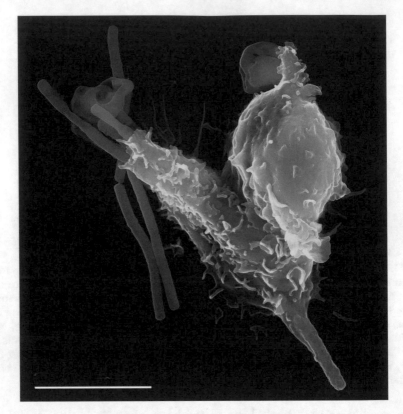

Fig. 11.2 A scanning electron microscope can take images that would be impossible using a traditional optical microscope. This image, for example, shows a white blood cell (yellow) engulfing anthrax bacteria (orange). The white line at the bottom indicates the scale of the image: the bar is 5 micrometers long. SEM and TEM images clearly demonstrate that the world described in "The Diamond Lens" does not exist at these distance scales (Credit: Volker Brinkmann)

* * *

Even a transmission electron microscope is limited in the magnification it can achieve. The world contains structures that are much smaller than can be seen by a TEM. Could there be creatures of interest existing at a scale that's hidden even from the prying eyes of an electron microscope?

Again, the answer is: no.

One of the main strands in the story of twentieth century physics is the development of our understanding of the world as it appears on the very smallest scales. The story began, as much of twentieth century physics began, in 1905—when Albert Einstein solved a puzzle first highlighted by the botanist Robert Brown in 1827. Brown used a microscope to study tiny granules inside pollen grains suspended in water. He noticed the granules were in constant, random motion. Similar random motion was later seen to be exhibited by microscopic chips of glass and scraps of stone, and you can observe the same effect if you watch dust motes dancing in a shaft of light, so the explanation for the motion clearly does not lie in biology. Einstein explained how Brownian motion could be explained if the pollen grains were being buffeted by water molecules: it was the first convincing evidence for a long-standing but controversial idea—that matter consists ultimately of molecules and atoms.

The word "atom" comes from a Greek word meaning "indivisible". But just a few years after Einstein's seminal paper on Brownian motion had started to persuade the world of the existence of atoms, physicists learned how to divide the atom. Ernest Rutherford demonstrated that atoms have a structure. He showed that an atom is mainly empty space: it has a central nucleus, which contains nearly all of an atom's mass, while a cloud of electrons orbit the nucleus at a distance.

The atomic nucleus also possesses a structure: it consists of particles called protons and neutrons. And just as the atom itself can be divided, so too can the atomic nucleus. In 1932, John Cockcroft and Ernest Walton succeeded in splitting the nucleus of a lithium atom by bombarding it with high energy particles. Over the following decades physicists continued with this approach—smashing particles together at ever greater energies and examining the debris—and it proved to be an exceptionally fruitful line of attack for investigating the subatomic world. It has been so successful that we now know—to some level of accuracy, at least—how the universe is put together. At the most fundamental level there are particles called electrons and quarks. Certain combinations of quarks are permitted and these allowed combinations form particles such as protons and neutrons. Protons and neutrons combine to form atomic nuclei. Atomic nuclei, when orbited by electrons, are the constituents of atoms. Atoms combine to form molecules. And molecules combine to form the complex structures we see around us—from microscopic organisms all the way up to macroscopic entities. Whether physicists are using a machine such as CERN's Large Hadron Collider to investigate the properties of quarks, or an advanced atomic force microscope to investigate matter at the nanoscale, they are always able to explain their observations in terms of standard models of physics. Nothing has been discovered, at any distance scale, that would permit the existence of the world inhabited by Animula.

Quite by accident, however, one aspect of O'Brien's story does strike a chord with modern readers: the title, "The Diamond Lens". The UK possesses a national research facility called a synchrotron—a huge ring, 500 m in circumference, which accelerates electrons in such a way that intense beams of light are created. The synchrotron acts as a giant microscope, which scientists use to study fossils, and oil paintings, and archeological artefacts, and viruses, and jet engines … it's an impressive operation. And the name of this giant microscope? Diamond!

* * *

For the final chapter we turn from speculations about what we might see through a microscope to speculations about what—if we're unlucky—we might one day see through a telescope.

Notes and Further Reading

For example, Antonie Philips van Leeuwenhoek—A quick online search will throw up numerous sites devoted to the history of microscopy. Rochow and Tucker (1994) is expensive, but if your library has it then you'll find a good overview of the development of microscopy. Asimov (1964) contains potted biographies of many of the scientists mentioned in this chapter.

to observe various types of specimen—Allen (2015) provides a concise account of what can be seen at different levels of magnification and of the various types of instrument.

appear in science fiction from time to time—"The Diamond Lens" was published in Victorian times (O'Brien 1858); *The Shrinking Man* was published almost a century later (Matheson 1956); and such stories continued—Asimov (1966), who was of course aware of the impossibility of miniaturization, nevertheless tried to provide a veneer of scientific respectability to this idea in his novelization of the film *Fantastic Voyage*.

microscopists have developed other techniques—The Nobel Prize committee has recognised various developments in microscopy over the decades, and their website gives a good account of techniques such as transmission electron microscopy. See Nobelprize.org (2014).

much of twentieth century physics began, in 1905—In addition to publishing papers on special relativity, mass–energy equivalence, and the photoelectric effect (for which he was awarded the Nobel prize), this miraculous year saw Einstein publish his explanation of Brownian motion. See Einstein (1905) if you want to read the original paper.

how the universe is put together—The Standard Model of particle physics gives us our best understanding of the subatomic world, and represents one of the pinnacles of human intellectual endeavor. Oerter (2006) gives a good overview of the historical development of the Standard Model, and is only slightly out of date. Carroll (2013), with its description of the hunt for the Higgs particle, brings the story up to date.

Bibliography

Allen, T.: Microscopy: A Very Short Introduction. Oxford University Press, Oxford (2015)

Asimov, I.: Asimov's Biographical Encyclopedia of Science and Technology. Doubleday, New York (1964)

Asimov, I.: Fantastic Voyage. Bantam, New York (1966)

Carroll, S.: The Particle at the End of the Universe: How the Hunt for the Higgs Boson Leads Us to the Edge of a New World. Dutton, New York (2013)

Einstein, A.: Über die von der molekularkinetischen Theorie der Wärme geforderte Bewegung von in ruhenden Flüssigkeiten suspendierten Teilchen. Ann. Phys. **17**(8), 549–560 (1905)

Matheson, R.: The Shrinking Man. Gold Medal Books, New York (1956)

Nobelprize.org: Microscopes. https://www.nobelprize.org/educational/physics/microscopes/ (2014)

O'Brien, F.J.: The diamond lens. Atlantic Month. **1**(3), 354–367 (1858)

Oerter, R.: The Theory of Almost Everything: The Standard Model, the Unsung Triumph of Modern Physics. Plume, New York (2006)

Rochow, T.G., Tucker, P.A.: Introduction to Microscopy by Means of Light, Electrons, X Rays, or Acoustics. Springer, Berlin (1994)

12

Impact Events

The Star (H.G. Wells)

It was on the first day of the New Year that the announcement was made, almost simultaneously from three observatories, that the motion of the planet Neptune, the outermost of all the planets that wheel about the sun, had become very erratic. Ogilvy had already called attention to a suspected retardation in its velocity in December. Such a piece of news was scarcely calculated to interest a world the greater portion of whose inhabitants were unaware of the existence of the planet Neptune, nor outside the astronomical profession did the subsequent discovery of a faint remote speck of light in the region of the perturbed planet cause any very great excitement. Scientific people, however, found the intelligence remarkable enough, even before it became known that the new body was rapidly growing larger and brighter, that its motion was quite different from the orderly progress of the planets, and that the deflection of Neptune and its satellite was becoming now of an unprecedented kind.

Few people without a training in science can realise the huge isolation of the solar system. The sun with its specks of planets, its dust of planetoids, and its impalpable comets, swims in a vacant immensity that almost defeats the imagination. Beyond the orbit of Neptune there is space, vacant so far as human observation has penetrated, without warmth or light or sound, blank emptiness, for twenty million times a million miles. That is the smallest

© Springer Nature Switzerland AG 2019
S. Webb, *New Light Through Old Windows: Exploring Contemporary Science Through 12 Classic Science Fiction Tales*, Science and Fiction,
https://doi.org/10.1007/978-3-030-03195-4_12

estimate of the distance to be traversed before the very nearest of the stars is attained. And, saving a few comets more unsubstantial than the thinnest flame, no matter had ever to human knowledge crossed this gulf of space, until early in the twentieth century this strange wanderer appeared. A vast mass of matter it was, bulky, heavy, rushing without warning out of the black mystery of the sky into the radiance of the sun. By the second day it was clearly visible to any decent instrument, as a speck with a barely sensible diameter, in the constellation Leo near Regulus. In a little while an opera glass could attain it.

On the third day of the new year the newspaper readers of two hemispheres were made aware for the first time of the real importance of this unusual apparition in the heavens. "A Planetary Collision," one London paper headed the news, and proclaimed Duchaine's opinion that this strange new planet would probably collide with Neptune. The leader writers enlarged upon the topic; so that in most of the capitals of the world, on January 3rd, there was an expectation, however vague, of some imminent phenomenon in the sky; and as the night followed the sunset round the globe, thousands of men turned their eyes skyward to see—the old familiar stars just as they had always been. Until it was dawn in London and Pollux setting and the stars overhead grown pale. The Winter's dawn it was, a sickly filtering accumulation of daylight, and the light of gas and candles shone yellow in the windows to show where people were astir. But the yawning policeman saw the thing, the busy crowds in the markets stopped agape, workmen going to their work betimes, milkmen, the drivers of news-carts, dissipation going home jaded and pale, homeless wanderers, sentinels on their beats, and in the country, labourers trudging afield, poachers slinking home, all over the dusky quickening country it could be seen—and out at sea by seamen watching for the day—a great white star, come suddenly into the westward sky!

Brighter it was than any star in our skies; brighter than the evening star at its brightest. It still glowed out white and large, no mere twinkling spot of light, but a small round clear shining disc, an hour after the day had come. And where science has not reached, men stared and feared, telling one another of the wars and pestilences that are foreshadowed by these fiery signs in the Heavens. Sturdy Boers, dusky Hottentots, Gold Coast Negroes, Frenchmen, Spaniards, Portuguese, stood in the warmth of the sunrise watching the setting of this strange new star.

And in a hundred observatories there had been suppressed excitement, rising almost to shouting pitch, as the two remote bodies had rushed together; and a hurrying to and fro, to gather photographic apparatus and spectroscope, and this appliance and that, to record this novel astonishing sight, the destruction of a world. For it was a world, a sister planet of our earth, far greater than our earth indeed, that had so suddenly flashed into flaming death. Neptune it was, had been struck, fairly and squarely, by the strange planet from outer space and the heat of the concussion had incontinently turned two solid globes into one vast mass of incandescence. Round the world that day, two hours before the dawn, went the pallid great white star, fading only as it sank westward and the sun mounted above it. Everywhere men marvelled at it, but of all those who saw it none could have marvelled more than those sailors, habitual watchers of the stars, who far away at sea had heard nothing of its advent and saw it now rise like a pigmy moon and climb zenithward and hang overhead and sink westward with the passing of the night.

And when next it rose over Europe everywhere were crowds of watchers on hilly slopes, on house-roofs, in open spaces, staring eastward for the rising of the great new star. It rose with a white glow in front of it, like the glare of a white fire, and those who had seen it come into existence the night before cried out at the sight of it. "It is larger," they cried. "It is brighter!" And, indeed the moon a quarter full and sinking in the west was in its apparent size beyond comparison, but scarcely in all its breadth had it as much brightness now as the little circle of the strange new star.

"It is brighter!" cried the people clustering in the streets. But in the dim observatories the watchers held their breath and peered at one another. "*It is nearer,*" they said. "*Nearer!*"

And voice after voice repeated, "It is nearer," and the clicking telegraph took that up, and it trembled along telephone wires, and in a thousand cities grimy compositors fingered the type. "It is nearer." Men writing in offices, struck with a strange realisation, flung down their pens, men talking in a thousand places suddenly came upon a grotesque possibility in those words, "It is nearer." It hurried along wakening streets, it was shouted down the frost-stilled ways of quiet villages; men who had read these things from the throbbing tape stood in yellow-lit doorways shouting the news to the passersby. "It is nearer." Pretty women, flushed and glittering, heard the news told jestingly between the dances, and feigned an intelligent interest they did not feel. "Nearer! Indeed. How curious! How very, very clever people must be to find out things like that!"

Lonely tramps faring through the wintry night murmured those words to comfort themselves—looking skyward. "It has need to be nearer, for the night's as cold as charity. Don't seem much warmth from it if it *is* nearer, all the same."

"What is a new star to me?" cried the weeping woman kneeling beside her dead.

The schoolboy, rising early for his examination work, puzzled it out for himself—with the great white star shining broad and bright through the frost-flowers of his window. "Centrifugal, centripetal," he said, with his chin on his fist. "Stop a planet in its flight, rob it of its centrifugal force, what then? Centripetal has it, and down it falls into the sun! And this—!

"Do *we* come in the way? I wonder—"

The light of that day went the way of its brethren, and with the later watches of the frosty darkness rose the strange star again. And it was now so bright that the waxing moon seemed but a pale yellow ghost of itself, hanging huge in the sunset. In a South African City a great man had married, and the streets were alight to welcome his return with his bride. "Even the skies have illuminated," said the flatterer. Under Capricorn, two negro lovers, daring the wild beasts and evil spirits, for love of one another, crouched together in a cane brake where the fire-flies hovered. "That is our star," they whispered, and felt strangely comforted by the sweet brilliance of its light.

The master mathematician sat in his private room and pushed the papers from him. His calculations were already finished. In a small white phial there still remained a little of the drug that had kept him awake and active for four long nights. Each day, serene, explicit, patient as ever, he had given his lecture to his students, and then had come back at once to this momentous calculation. His face was grave, a little drawn and hectic from his drugged activity. For some time he seemed lost in thought. Then he went to the window, and the blind went up with a click. Half way up the sky, over the clustering roofs, chimneys and steeples of the city, hung the star.

He looked at it as one might look into the eyes of a brave enemy. "You may kill me," he said after a silence. "But I can hold you—and all the universe for that matter—in the grip of this little brain. I would not change. Even now."

He looked at the little phial. "There will be no need of sleep again," he said. The next day at noon—punctual to the minute, he entered his lecture theatre, put his hat on the end of the table as his habit was, and carefully selected a large piece of chalk. It was a joke among his students that he could not lecture without that piece of chalk to fumble in his fingers, and once he had been stricken to impotence by their hiding his supply. He came and looked under his grey eyebrows at the rising tiers of young fresh faces, and spoke with his accustomed studied commonness of phrasing. "Circumstances have arisen— circumstances beyond my control," he said and paused, "which will debar me from completing the course I had designed. It would seem, gentlemen, if I may put the thing clearly and briefly, that—Man has lived in vain."

The students glanced at one another. Had they heard aright? Mad? Raised eyebrows and grinning lips there were, but one or two faces remained intent upon his calm grey-fringed face. "It will be interesting," he was saying, "to devote this morning to an exposition, so far as I can make it clear to you, of the calculations that have led me to this conclusion. Let us assume—"

He turned towards the blackboard, meditating a diagram in the way that was usual to him. "What was that about 'lived in vain?'" whispered one student to another. "Listen," said the other, nodding towards the lecturer.

And presently they began to understand.

That night the star rose later, for its proper eastward motion had carried it some way across Leo towards Virgo, and its brightness was so great that the sky became a luminous blue as it rose, and every star was hidden in its turn, save only Jupiter near the zenith, Capella, Aldebaran, Sirius and the pointers of the Bear. It was very white and beautiful. In many parts of the world that night a pallid halo encircled it about. It was perceptibly larger; in the clear refractive sky of the tropics it seemed as if it were nearly a quarter the size of the moon. The frost was still on the ground in England, but the world was as brightly lit as if it were midsummer moonlight. One could see to read quite ordinary print by that cold clear light, and in the cities the lamps burnt yellow and wan.

And everywhere the world was awake that night, and throughout Christendom a sombre murmur hung in the keen air over the country side like the belling of bees in the heather, and this murmurous tumult grew to a clangour in the cities. It was the tolling of the bells in a million belfry towers and steeples, summoning the people to sleep no more, to sin no more, but to gather in their churches and pray. And overhead, growing larger and brighter as the earth rolled on its way and the night passed, rose the dazzling star.

And the streets and houses were alight in all the cities, the shipyards glared, and whatever roads led to high country were lit and crowded all night long. And in all the seas about the civilised lands, ships with throbbing engines, and ships with bellying sails, crowded with men and living creatures, were standing out to ocean and the north. For already the warning of the master mathematician had been telegraphed all over the world, and translated into a hundred tongues. The new planet and Neptune, locked in a fiery embrace, were whirling headlong, ever faster and faster towards the sun. Already every second this blazing mass flew a hundred miles, and every second its terrific velocity increased. As it flew now, indeed, it must pass a hundred million of miles wide of the earth and scarcely affect it. But near its destined path, as yet only slightly perturbed, spun the mighty planet Jupiter and his moons sweeping splendid round the sun. Every moment now the attraction between the fiery star and the greatest of the planets grew stronger. And the result of that attraction? Inevitably Jupiter would be deflected from its orbit into an elliptical path, and the burning star, swung by his attraction wide of its sunward rush, would "describe a curved path" and perhaps collide with, and certainly pass very close to, our earth. "Earthquakes, volcanic outbreaks, cyclones, sea waves, floods, and a steady rise in temperature to I know not what limit"—so prophesied the master mathematician.

And overhead, to carry out his words, lonely and cold and livid, blazed the star of the coming doom.

To many who stared at it that night until their eyes ached, it seemed that it was visibly approaching. And that night, too, the weather changed, and the frost that had gripped all Central Europe and France and England softened towards a thaw.

But you must not imagine because I have spoken of people praying through the night and people going aboard ships and people fleeing toward mountainous country that the whole world was already in a terror because of the star. As a matter of fact, use and wont still ruled the world, and save for the talk of idle moments and the splendour of the night, nine human beings out of ten were still busy at their common occupations. In all the cities the shops, save one here and there, opened and closed at their proper hours, the doctor and the undertaker plied their trades, the workers gathered in the factories, soldiers drilled, scholars studied, lovers sought one another, thieves lurked and fled, politicians planned their schemes. The presses of the newspapers roared through the night, and many a priest of this church and that would not open his holy building to further what he considered a foolish panic. The newspapers insisted on the lesson of the year 1000—for then, too, people had anticipated the end. The star was no star—mere gas—a comet; and were it a star it could not possibly strike the earth. There was no precedent for such a thing. Common sense was sturdy everywhere, scornful, jesting, a little inclined to persecute the obdurate fearful. That night, at seven-fifteen by Greenwich time, the star would be at its nearest to Jupiter. Then the world would see the turn things would take. The master mathematician's grim warnings were treated by many as so much mere elaborate self-advertisement. Common sense at last, a little heated by argument, signified its unalterable convictions by going to bed. So, too, barbarism and savagery, already tired of the novelty, went about their nightly business, and save for a howling dog here and there, the beast world left the star unheeded.

And yet, when at last the watchers in the European States saw the star rise, an hour later it is true, but no larger than it had been the night before, there were still plenty awake to laugh at the master mathematician—to take the danger as if it had passed.

But hereafter the laughter ceased. The star grew—it grew with a terrible steadiness hour after hour, a little larger each hour, a little nearer the midnight zenith, and brighter and brighter, until it had turned night into a second day. Had it come straight to the earth instead of in a curved path, had it lost no velocity to Jupiter, it must have leapt the intervening gulf in a day, but as it was it took five days altogether to come by our planet. The next night it had become a third the size of the moon before it set to English eyes, and the thaw was assured. It rose over America near the size of the moon, but blinding white to look at, and *hot*; and a breath of hot wind blew now with its rising and gathering strength, and in Virginia, and Brazil, and down the St. Lawrence valley, it shone intermittently through a driving reek of thunder-clouds, flickering violet lightning, and hail unprecedented. In Manitoba was a thaw and devastating floods. And upon all the mountains of the earth the snow and ice began to melt that night, and all the rivers coming out of high country flowed thick and turbid, and soon—in their upper reaches—with swirling trees and the bodies of beasts and men. They rose steadily, steadily in the ghostly brilliance, and came trickling over their banks at last, behind the flying population of their valleys.

And along the coast of Argentina and up the South Atlantic the tides were higher than had ever been in the memory of man, and the storms drove the waters in many cases scores of miles inland, drowning whole cities. And so great grew the heat during the night that the rising of the sun was like the coming of a shadow. The earthquakes began and grew until all down America from the Arctic Circle to Cape Horn, hillsides were sliding, fissures were opening, and houses and walls crumbling to destruction. The whole side of Cotopaxi slipped out in one vast convulsion, and a tumult of lava poured out so high and broad and swift and liquid that in one day it reached the sea.

So the star, with the wan moon in its wake, marched across the Pacific, trailed the thunderstorms like the hem of a robe, and the growing tidal wave that toiled behind it, frothing and eager, poured over island and island and swept them clear of men. Until that wave came at last—in a blinding light and with the breath of a furnace, swift and terrible it came—a wall of water, fifty feet high, roaring hungrily, upon the long coasts of Asia, and swept inland across the plains of China. For a space the star, hotter now and larger and brighter than the sun in its strength, showed with pitiless brilliance the wide and populous country; towns and villages with their pagodas and trees, roads, wide cultivated fields, millions of sleepless people staring in helpless terror at the incandescent sky; and then, low and growing, came the murmur of the flood. And thus it was with millions of men that night—a flight nowhither, with limbs heavy with heat and breath fierce and scant, and the flood like a wall swift and white behind. And then death.

China was lit glowing white, but over Japan and Java and all the islands of Eastern Asia the great star was a ball of dull red fire because of the steam and smoke and ashes the volcanoes were spouting forth to salute its coming. Above was the lava, hot gases and ash, and below the seething floods, and the whole earth swayed and rumbled with the earthquake shocks. Soon the immemorial snows of Thibet and the Himalaya were melting and pouring down by ten million deepening converging channels upon the plains of Burmah and Hindostan. The tangled summits of the Indian jungles were aflame in a thousand places, and below the hurrying waters around the stems were dark objects that still struggled feebly and reflected the blood-red tongues of fire. And in a rudderless confusion a multitude of men and women fled down the broad river-ways to that one last hope of men—the open sea.

Larger grew the star, and larger, hotter, and brighter with a terrible swiftness now. The tropical ocean had lost its phosphorescence, and the whirling steam rose in ghostly wreaths from the black waves that plunged incessantly, speckled with storm-tossed ships.

And then came a wonder. It seemed to those who in Europe watched for the rising of the star that the world must have ceased its rotation. In a thousand open spaces of down and upland the people who had fled thither from the floods and the falling houses and sliding slopes of hill watched for that rising in vain. Hour followed hour through a terrible suspense, and the star rose not. Once again men set their eyes upon the old constellations they had counted lost to them forever. In England it was hot and clear overhead, though the ground quivered perpetually, but in the tropics, Sirius and Capella and Aldebaran showed through a veil of steam. And when at last the great star rose near ten hours late, the sun rose close upon it, and in the centre of its white heart was a disc of black.

Over Asia it was the star had begun to fall behind the movement of the sky, and then suddenly, as it hung over India, its light had been veiled. All the plain of India from the mouth of the Indus to the mouths of the Ganges was a shallow waste of shining water that night, out of which rose temples and palaces, mounds and hills, black with people. Every minaret was a clustering mass of people, who fell one by one into the turbid waters, as heat and terror overcame them. The whole land seemed a-wailing and suddenly there swept a shadow across that furnace of despair, and a breath of cold wind, and a gathering of clouds, out of the cooling air. Men looking up, near blinded, at the star, saw that a black disc was creeping across the light. It was the moon, coming between the star and the earth. And even as men cried to God at this respite, out of the East with a strange inexplicable swiftness sprang the sun. And then star, sun and moon rushed together across the heavens.

So it was that presently, to the European watchers, star and sun rose close upon each other, drove headlong for a space and then slower, and at last came to rest, star and sun merged into one glare of flame at the zenith of the sky. The moon no longer eclipsed the star but was lost to sight in the brilliance of the sky. And though those who were still alive regarded it for the most part with that dull stupidity that hunger, fatigue, heat and despair engender, there were still men who could perceive the meaning of these signs. Star and earth had been at their nearest, had swung about one another, and the star had passed. Already it was receding, swifter and swifter, in the last stage of its headlong journey downward into the sun.

And then the clouds gathered, blotting out the vision of the sky, the thunder and lightning wove a garment round the world; all over the earth was such a downpour of rain as men had never before seen, and where the volcanoes flared red against the cloud canopy there descended torrents of mud. Everywhere the waters were pouring off the land, leaving mud-silted ruins, and the earth littered like a storm-worn beach with all that had floated, and the dead bodies of the men and brutes, its children. For days the water streamed off the land, sweeping away soil and trees and houses in the way, and piling huge dykes and scooping out Titanic gullies over the country side. Those were the days of darkness that followed the star and the heat. All through them, and for many weeks and months, the earthquakes continued. But the star had passed, and men, hunger-driven and gathering courage only slowly, might creep back to their ruined cities, buried granaries, and sodden fields. Such few ships as had escaped the storms of that time came stunned and shattered and sounding their way cautiously through the new marks and shoals of once familiar ports. And as the storms subsided men perceived that everywhere the days were hotter than of yore, and the sun larger, and the moon, shrunk to a third of its former size, took now fourscore days between its new and new.

But of the new brotherhood that grew presently among men, of the saving of laws and books and machines, of the strange change that had come over Iceland and Greenland and the shores of Baffin's Bay, so that the sailors coming there presently found them green and gracious, and could scarce believe their eyes, this story does not tell. Nor of the movement of mankind now that the earth was hotter, northward and southward towards the poles of the earth. It concerns itself only with the coming and the passing of the Star. The Martian astronomers—for there are astronomers on Mars, although they are very different beings from men—were naturally profoundly interested by these things. They saw them from their own standpoint of course. "Considering the mass and temperature of the missile that was flung through our solar system into the sun," one wrote, "it is astonishing what a little damage the earth, which it missed so narrowly, has sustained. All the familiar continental markings and the masses of the seas remain intact, and indeed the only difference seems to be a shrinkage of the white discoloration (supposed to be frozen water) round either pole." Which only shows how small the vastest of human catastrophes may seem, at a distance of a few million miles.

> **Herbert George Wells** (1866–1946) is undoubtedly one of the most influential SF authors of all time. Several of his novels—including *The Time Machine* (1895), *The Island of Doctor Moreau* (1896), *The Invisible Man* (1897), and *The War of the Worlds* (1898)—are classics of the field. In addition to his science fiction Wells wrote mainstream fiction as well as treatises on history, politics, and social affairs; he was nominated four times for the Nobel prize in literature. But it is as an SF author that he remains best known.

Commentary

H.G. Wells pioneered many science fiction themes: time travel, invisibility, malevolent aliens—and, in this story, impact events. The science in "The Star" has dated but the story's core message—that a celestial body could cause catastrophe here on Earth—is even more plausible now than it was when Wells wrote his tale in 1897.

In terms of the astronomical knowledge of his day, the overall setting Wells describes in "The Star" is essentially correct. Consider, for example, his description of the outer solar system. He was right in saying the outermost planet is Neptune: the planet had been discovered half a century earlier, in 1846. Furthermore, just 17 days after Neptune was first identified, astronomers discovered it had a satellite. We now call the satellite Triton, but that name didn't come into common parlance until the 1930s. Up until then scientists referred to it in the same way Wells refers to it: "the satellite of Neptune". There are now 14 known satellites of Neptune, but it wasn't until 1949 that astronomers found a second Neptunian moon. Wells was faithful to the science as it then stood.

Or consider the description Wells gives of stellar distances. By 1897, the distance to about sixty stars had been determined using the method of parallactic shift. So Wells, in his story, was thus giving an accurate feel for the immense distances separating the stars.

Wells was less convincing in his description of the effects of the trespassing star on the orbital dynamics of the solar system. He was also far from convincing about the nature of the star itself. Indeed, in some places in the story it's not clear whether Wells thinks of the interloper as a star or a planet. Whatever Wells imagined, though, we now understand that no star is going to barrel in from outer space and disrupt the solar system. Most stars move through space with a velocity, relative to the Sun, of a few kilometres per second. At this speed, even if the nearest star happened to be on a collision course with the Sun (and it isn't) then we'd be waiting for a million years for the crash to happen. We wouldn't be taken by surprise. It's true that astronomers have identified a few so-called hypervelocity stars, which move at speeds of up to 1000 km/s, but even if the nearest star were a hypervelocity star (which it isn't) and heading straight for us (which it isn't) there'd still be a delay of one thousand years before it reached us. The notion that a star could appear seemingly from nowhere and then, within a few days, disrupt the solar system is not credible.

Nevertheless, although his description of a wandering star disturbing the clockwork mechanism of our solar system was wrong, Wells was right to highlight the possibility of life on Earth being disrupted by celestial bodies. And in this he proved himself to be farsighted. The science of his day, which was coming to terms with the implications of Darwin's theory of evolution, was emphasising gradual—almost imperceptible—change. The notion of catastrophic change was frowned upon. But we now know catastrophe can indeed come from the skies.

About ten years after Wells published "The Star", a mysterious event occurred near the Stony Tunguska River in Siberia. An explosion flattened more than two thousand square kilometres of forest (see Fig. 12.1). Few people lived in Tunguska, so no casualties were reported; had the same event occurred in London or New York, millions of people might have died. The explosion was an impact event—the air burst of a meteoroid. The meteoroid disintegrated before it hit Earth's surface, so it's difficult to know for sure what the object was. The best guesses, however, are either that it was an asteroid about 60 metres in diameter or a comet about 190 metres in diameter.

The Tunguska event itself might not have been a catastrophic occurrence, but over recent decades scientists have come to appreciate that similar, much larger, events have influenced the development of life on Earth.

Fig. 12.1 A photograph taken by a member of the 1929 expedition to investigate the Tunguska event. The explosion flattened vast swathes of forest: it's estimated that 80 million trees were toppled. The energy release involved in this event was much greater than when the first atom bomb exploded over Hiroshima (Credit: Public domain)

* * *

Buried beneath Mexico's Yucatán Peninsula is a crater called Chicxulub, which gets its name from a nearby town. The crater was the result of an impact event, a collision involving an asteroid or comet about 10 to 15 kilometres in diameter. The Chicxulub impactor was *much* bigger than the Tunguska impactor, and the effects were bigger too. As with Tunguska, no people died when the meteoroid hit Chicxulub—but only because no humans were around at the time of the impact. The collision happened about 66 million years ago, long before humans were on the scene. The dominant animals back then were dinosaurs, and the worldwide climate disruption caused by the impact helped them on their way to extinction. Indeed, Chicxulub caused a mass extinction event: about three quarters of all plant and animal species died out. This was a clear case of destruction raining down from the heavens. If a similar sized meteoroid struck Earth today, humans would almost certainly suffer the same fate as the creatures who were around when the Chicxulub impactor struck.

We now know that the history of life on Earth is not just the story of natural selection operating gradually and infinitesimally slowly over long aeons. Our planet gets peppered with space rocks—most of them small; a few of them large; an occasional monster—and these impacts have the capacity to alter the course of evolution. Indeed, in 1984 an influential paper claimed that, if one examines the fossil record in detail, it's possible to discern a periodicity in mass extinction events: they seem to occur at intervals of about 26 million years. Two groups of astronomers then postulated a mechanism that could explain such a cycle. Suppose the Sun had a companion—a small red dwarf or brown dwarf star, orbiting the Sun at a distance of about 1.5 light years. Such a companion would exert a gravitational tug on the Oort Cloud of comets, disrupting orbits and increasing the number of comets falling into the inner Solar System—and thus increasing the likelihood of an impact event on Earth. This hypothetical companion star was given the name Nemesis.

Numerous surveys have failed to detect the existence of any Nemesis-like object orbiting the Sun, and more recent analyses of the fossil record suggest the claimed periodicity in mass extinction events was a statistical artefact. But the matter isn't entirely settled—and so it's interesting to speculate that a star might well be the cause of death and destruction here on Earth; not a wandering star, as Wells envisaged, but a constant companion star to the Sun.

* * *

Science has updated another element of the "The Star". Wells, when discussing the vastness of the cosmos, stated that "no matter had ever to human knowledge crossed this gulf of space". Well, that was true when Wells was writing. But in 1911 Victor Hess began a series of experiments that led to the discovery of cosmic rays—particles from space that bombard Earth's atmosphere. We now know that some cosmic rays can originate from far outside the solar system; indeed, some of the highest energy cosmic rays originate from outside our Milky Way galaxy. And in November 2017, as I began to re-read Wells' story for this volume, astronomers provided details of a visitor from another solar system—not a mere subatomic particle but a piece of rock. A rock that had indeed crossed the "gulf of space".

The International Astronomical Union called the object 1 I/2017 U1; you can also refer to it as 'Oumuamua, which means "scout" in Hawaiian. No object like 'Oumuamua had ever been seen before in the solar system. It is relatively small, extremely dark, and unremarkable except for its highly elongated shape (it's a spear of rock about 400 metres long, very much along the lines of how I visualise the mysterious craft in Arthur Clarke's *Rendezvous with Rama*; see Fig. 12.2). What sets 'Oumuamua apart is its velocity. It's moving so fast, relative to the Sun, that it can't have originated in the solar system and

Fig. 12.2 An artist's impression of the first detected interstellar asteroid, which was discovered on 19 October 2017 by astronomers using the Pan-STARRS 1 telescope in Hawaii. The spear-shaped object is dark, about 400 metres long, and was wandering through space for millions of years before its chance encounter with our Solar System (Credit: ESO/M. Kornmesser)

it won't be captured by the solar system. 'Oumuamua came from a distant planetary system (it's not known which) and it will soon leave our solar system. Although 'Oumuamua is the first such interloper to be found, some estimates suggest that on average three such objects enter our solar system each day and three leave each day. Fortunately, 'Oumuamua is not on a collision course with Earth. But if a similar object *were* to strike Earth, the damage to our civilisation might be worse than anything Wells described. And we would have much less time to prepare for our end than the characters did in "The Star": the dark red surface of 'Oumuamua absorbs 96% of the light that falls upon it, and if that is typical of these interstellar interlopers then we'd have a hard time spotting the object before it struck.

We shouldn't be surprised that Earth is subject to impacts: look up at the Moon and you see a surface pocked by innumerable craters, the result of countless impacts that have happened over the past four billion years. The Moon isn't some sort of magnet for incoming meteoroids. The reason we see so many craters on the Moon is because our satellite has no weather systems to erode them, no system of plate tectonics to erase them. Earth is struck by meteoroids with the same frequency as the Moon (indeed, we believe the Moon itself was created when Earth was hit by a Mars-sized object) but Earth's various systems tend to expunge the resulting impact craters.

Meteoroids have struck Earth in the past and they'll strike Earth in the future. Indeed, if you consider the statistics, you're more likely to die from a meteor strike than you are from a lightning strike. There isn't much chance of a large asteroid dropping on us any time soon—but if one *did* impact then all seven billion of us would die. Multiply the small chance of a strike happening with the total devastation caused by such an impact and you get a result that is far from negligible. We insure against many events that are unlikely to happen; surely our civilisation could develop a better guard against asteroid strikes? It would be money well spent.

To end on a truly depressing note: the cosmos offers existential threats against which we have no control. Consider, for example, a supernova—the final explosion of a high-mass star. A supernova releases vast amounts of radiation. If a nearby star went supernova then many life forms here on Earth would be threatened. Indeed, some scientists have suggested that supernovae may have been responsible for at least some of Earth's past mass extinctions. Or consider a gamma-ray burst—the most powerful sort of explosions in the universe, which can result from either a neutron star merger or the collapse of a rapidly rotating, high-mass star. If Earth found itself in the jet of radiation coming from a nearby gamma-ray burst then the planet would be toast. If you are particularly neurotic then you might also want to worry about the possibility of false vacuum collapse. Our universe *might* be stuck in a situation in which the vacuum is not in its lowest energy state. If that's the case the universe could tunnel from the false vacuum to a state of lower energy. But there's little point in fretting about this particular threat. If the false vacuum did collapse then it's entirely possible that all matter would be destroyed instantaneously and without warning. You wouldn't feel a thing.

Notes and Further Reading

the outermost planet is Neptune—For details of the discovery of Neptune see, for example, Standage (2000).

the method of parallactic shift—Measuring a stellar parallax involves carefully recording the position of the star, relative to the background of distant fixed stars. If the star's position is observed in January and July, say, then that six-month separation means the star is being observed from vantage points separated by the diameter of Earth's orbit around the Sun. If the star is nearby then it's apparent position will shift, just as a pencil held at arm's length shifts its position relative to the background when you observe it first with only your left eye closed and then with only your right eye closed.

From the size of the parallactic displacement of an object you can determine the distance to that object. See Webb (1999) for further details of this and other methods for measuring astronomical distances.

a mysterious event occurred near the Stony Tunguska River—I first learned of the Tunguska event from Baxter and Atkins (1975). This book is still worth reading, but Rubtsov (2009) gives a much more in-depth of the explosion and what might have caused it.

a crater called Chicxulub—For further details of the Chicxulub impact, and other impact events, see for example Verschuur (1997).

a periodicity in mass extinction events—The periodicity was claimed by Raup and Sepkoski (1984), who examined the fossil record and identified 12 extinction events over the past 250 million years.

a mechanism that could explain such a cycle—The Nemesis hypothesis was proposed independently by Whitmire and Jackson (1984) and Davis, Hut, and Muller (1984).

a statistical artefact—Bailer-Jones (2011) found no evidence for periodicity in Earth's impact record, and thus removed the need to hypothesise the existence of Nemesis. Furthermore, no sign of a companion star to the Sun as surfaced in any astronomical survey.

the object 1 I/2017 U1—For details of the initial discovery, see Meech et al. (2017). The resemblance of 'Oumuamua to the spaceship Rama, as described by Clarke (1973), made it inevitable that astronomers listened for possible alien signals! But nothing was heard.

a truly depressing note—For a fun guide to some possible routes to Armageddon, see Darling and Schulze-Makuch (2012).

Bibliography

Bailer-Jones, C.A.L.: Bayesian time series analysis of terrestrial impact cratering. Mon. Notes R. Astro. Soc. **416**(2), 1163–1180 (2011)

Baxter, J., Atkins, T.: The Fire Came By: The Riddle of the Great Siberian Explosion. Macdonald and Jane's, London (1975)

Clarke, A.C.: Rendezvous With Rama. Gollancz, London (1973)

Darling, D., Schulze-Makuch, D.: Megacatastrophes! Nine Strange Ways the World Could End. Oneworld, London (2012)

Davis, M., Hut, P., Muller, R.A.: Extinction of species by periodic comet showers. Nature. **308**(5961), 715–717 (1984)

Meech, K.J., et al.: A brief visit from a red and extremely elongated interstellar asteroid. Nature. (2017). https://doi.org/10.1038/nature25020

Raup, D.M., Sepkoski Jr., J.J.: Periodicities of extinctions in the geologic past. Proc. Natl. Acad. Sci. **81**(3), 801–805 (1984)

Rubtsov, V.: The Tunguska Mystery. Springer, Berlin (2009)

Standage, T.: The Neptune File: A Story of Astronomical Rivalry and the Pioneers of Planet Hunting. Walker, London (2000)

Verschuur, G.L.: Impact! The Threat of Comets and Asteroids. Oxford University Press, Oxford (1997)

Webb, S.: Measuring the Universe: The Cosmological Distance Ladder. Springer, Berlin (1999)

Wells, H.G.: The star. The Graphic. December issue (1897)

Whitmire, D.P., Jackson, A.A.: Are periodic mass extinctions driven by a distant solar companion? Nature. **308**(5961), 713–715 (1984)

Printed in the United States
By Bookmasters